Geological Time

CONTEMPORARY
SCIENCE
PAPERBACKS
46

2.4.73

J. F. KIRKALDY D.SC., F.G.S.
Professor of Geology, Queen Mary College, London

Geological Time

OLIVER & BOYD   *Edinburgh*

OLIVER AND BOYD
Tweeddale Court Edinburgh EH1 1YL
A Division of Longman Group Ltd

First published 1971
© J. F. Kirkaldy 1971

Set in Times New Roman and printed in
Great Britain by Cox & Wyman Ltd,
London, Fakenham and Reading

Preface

Geochronology is one of many growing points of modern geology. During the past twenty years techniques have been developed which enable us for the first time to date absolutely, that is in terms of millions of years, many of the events of a past stretching back to the formation of the Earth's crust more than four and a half thousand million years ago. I have attempted to trace the development of geochronological ideas; to assess, I trust realistically, our present state of knowledge, and to suggest some, at least, of the advances in knowledge that will be made in the future as techniques become more refined and, more important, we learn the real meaning of many of the events dated.

My colleagues, Drs A. C. Bishop and W. E. French have read the typescript and I should like to express my warm appreciation of their helpful criticisms.

J. F. KIRKALDY

Contents

1. Introduction

At their nearest points of approach, the Earth's satellite, the Moon, the planet Venus and the planet Mars are 221 463, 26 000 000 and 35 000 000 miles respectively distant from the Earth. Until a few years ago such figures were of little significance except to astronomers, but with the development of space craft, everyone is beginning to appreciate the dimensions of the nearer parts of the Solar System.

In the same way, good television programmes, well-written articles in the better magazines, many popular books on archaeological subjects, have made us familiar with the changing civilizations, architectural styles and manners of life of our race during the past few thousand years. Nevertheless, these cover but the veriest fraction of the span of time during which the planet we inhabit has existed. The Minoan civilization of Crete dates back to about 4000 years B.C. The Earth was formed about 4600 million years ago and man has inhabited its surface for only the last one million or so years of this vast period of time. To quote the American writer, W. Bascom, 'man's entire span is one page in a book of 50 000 pages, of which written history is only the last line'.

In this book we trace the ways in which our knowledge of Earth-history has been built up. It is now becoming possible to draw up a chronology dated in millions of years from the present. But this is quite a recent development: up till little more than a century and a half ago, not only was the Biblical story of the creation of the Earth in six days rigidly accepted, but it was also believed that this event took place in the year 4004 B.C. In the next chapter we trace in outline the work of the founders of the geological science. They

1

clearly appreciated that a mere six thousand years was far too short a span of time for all the events whose story they had begun to decipher by using the Doctrine of Uniformitarianism ('the present is the key to the past') to interpret what has been called the 'record of the rocks'. Geologists then hammered out their own chronology, the Stratigraphical Table, in which the various distinctive rock layers were arranged in the order in which they were formed. But with purely geological methods alone, it was impossible to erect other than a relative, though often very detailed, time-scale – this is older than that. There were so many unknowns that it was impossible to determine, save in a few very exceptional cases, the time-span in years needed for the formation of individual rock layers, such as the familiar Chalk which is exposed over so much of eastern England and the neighbouring parts of Europe.

Moreover, the Stratigraphical Table is based on the use of fossils for determining whether or not rock units are of the same or of different relative age. Attention was naturally concentrated on the rocks which usually contain fossils, in many places in great abundance and variety. But underlying, and therefore older than, the fossil-bearing Phanerozoic beds are the vast thicknesses of virtually unfossiliferous Pre-Cambrian rocks. It is only within the last two or three decades that the complex story of the Pre-Cambrian rocks has begun to be unravelled. Towards the close of the last century, geologists had deciphered a sufficiently long Earth-history to be confident that it represented a vast period of time extending back through hundreds of millions of years. But the physicist Lord Kelvin would not agree; attempting to calculate the age of the Earth on a basis of its rate of cooling from an original molten state, he insisted that geological time must be restricted to between 20 and 40 million years.

Lord Kelvin's calculations were made shortly before the discovery of the phenomenon of radioactivity. Not only did this provide a hitherto unknown source for heat in the rocks of the Earth's crust, but when the rate of radioactive disintegration was determined, it became possible to use radioactive minerals as geological clocks. The time of formation millions of years ago of a particular radioactive mineral could be calculated. If the geological age of the rocks containing the mineral was known, then the relative age of this point on the Stratigraphical Table could be converted into an absolute age in millions of years.

In Chapter 5 the various methods used in isotopic (radiometric) age-determination are described and their applicability discussed. But whilst isotopic age-determinations are a very considerable technological breakthrough, the results obtained are often far more difficult to interpret than was expected to be the case only a few years ago, when most of the modern age-determination laboratories were set up. Chapter 6 gives a critical discussion of the use that can be made at present of age-determinations, both in dating in years the time at which rock layers were deposited, and also the times at which mountain building episodes occurred. In particular, it will be shown how these new techniques have increased our knowledge of the events of Pre-Cambrian times, that is of the period earlier than about 600 million years ago.

Isotopic dating is one of the growing points of geology. An accurate time-scale would give a fourth dimension to many branches of the science, so in the final chapter we look ahead and consider some of the existing uncertainties and problems that it is hoped will be clarified as our knowledge of and practice in age-determination improves.

2. Appreciation of the Long Duration of Geological Time

The word geology is made up of the Greek roots *Gaia* or *Ge* (the mythical earth-goddess) and *logos* (knowledge of). Knowledge of the Earth includes not only its form and constitution, but also its history. The Earth is not unique, but is a part of one of the myriads of galaxies studied by astronomers. The events that have lead up to the creation of our particular galaxy, the Solar System, and to the creation of the Earth within that galaxy fall within the field of astronomy rather than of geology. The Earth is however unique, as far as we definitely know at present, in that the atmospheric and other conditions of its outer surface were suitable for the development of life. As we shall see later, the study of the rocks that make up the surface layers of the Earth, enables us to trace an exceedingly long and ever-changing sequence of different forms of life, culminating in the appearance of man. So geology, or rather that branch of the science known as stratigraphy or historical geology, grades backwards into astronomy and forwards into archaeology and then, when written records become freely available, into history.

The sages and philosophers of the ancient civilization of the East are known to have speculated on the duration of Earth-history, The Chaldeans, who had considerable astronomical knowledge, believed that the Earth had been in existence for 21 500 000 years, whilst the ancient Hindu philosophers had an even more precise chronology, according to which the earth was created 1 972 949 071 years before the present (A.D. 1970).

In the fragments of the writings of ancient Greek philosophers that have come down to us, there are accounts,

4

notably those by Theophrastus (*c.* 370–287 B.C.), of the localities and uses of certain rocks and minerals. Others have left descriptions of the effects of earthquakes and volcanic eruptions. The Roman, Pliny the Elder (A.D. 23–79) described in his *Natural History* many rocks and minerals. But these classical writers had no real conception of Earth-history, for they accepted the wildest speculations as to the origin of minerals and other objects found in the rocks, without subjecting them to the test of careful observation of the manner in which these minerals, etc., occurred.

From the Middle Ages until nearly the close of the eighteenth century European thought on Earth-history was dominated by the story in the first chapter of Genesis of the Creation of the Earth in six days. In 1644, John Lightfoot, a distinguished Greek scholar who was Vice-Chancellor of Cambridge University, calculated from the textual evidence that the moment of creation was nine o'clock in the morning of September 17th, 3928 B.C. Six years later Archbishop Ussher, Primate of Ireland, amended this to the entrance of the night preceding October 23rd, 4004 B.C. This date, 4004 B.C., was inserted by William Lloyd, Bishop of Winchester, in the Great (1701) Edition of the English Bible and it appeared as a marginal note in later editions of the Authorized Version. It was firmly accepted by theologians. In 1900 the Cambridge University Press stopped printing the note, followed ten years later by the Oxford University Press.

In 1749 Count de Buffon (1707–88), a distinguished French savant, suggested in the first volume of his *Natural History* that the six days of the Creation were six long periods of time, but he was forced by the power of the Church to recant and to declare that he accepted the Old Testament story as giving a complete and true history of the world.

With the Earth's history compressed into less than six thousand years, it was impossible to assume, as de Buffon and others had suggested, that topographic features such as mountains and river valleys had been produced slowly and gradually. On the contrary, according to the catastrophic viewpoint, river valleys were regarded as clefts opened by earthquakes, whilst mountains must have arisen very rapidly through great forces within the Earth acting with extreme violence. Noah's Flood must have been universal. The spreads of sand and gravel mantling so much of the surface of Europe were claimed as evidence of this.

But during the sixteenth, seventeenth and the first half of the eighteenth centuries, a few men had discovered, by careful observation, some of the principles on which our present knowledge of Earth-history is based. The fact that fossils are the remains of once living organisms and not of inorganic or of supernatural origin, was fully understood by that versatile genius Leonardo da Vinci (1452–1519) and by the Dane, Nicolaus Steno (1638–86). But in conformity with the religious thought of the time, Steno stated that the sea-shells which he had found in the rocks around Florence in Italy had been carried there by the Noachian Deluge. Steno also recognized that the rocks around Florence occurred as definite layers, each of which was traceable across the countryside, and that in such a series the higher bed must be the younger, for it must have been laid down on top of an earlier deposited layer. This concept, the Law of Super-position, is the basis on which the relative geological time-scale, described in the next chapter, has been erected. Steno's work, published in his *Prodromus* (1669), was unfortunately soon forgotten and did not make the impact on contemporary thought that it deserved.

The dogmatic catastrophic views of Earth-history were only overthrown after long continued and often extremely bitter controversy, which was initiated by the work of Dr James Hutton of Edinburgh (1726–97), who has been widely hailed as the Founder of Modern Geology. After a brief apprenticeship to law, which he soon abandoned, Hutton qualified as a doctor in 1749 after studying at Edinburgh, Paris and Leyden, but he never practised. Turning to agriculture, he quickly became interested in, and indeed one of the most successful practitioners of, the newly developed scientific methods of agriculture. By their use he brought his Berwickshire farm into a very high state of productivity. During these years he studied both the rocks underlying the lands which he and his friends farmed and also the effects of rain, frost, etc. In 1768 Hutton gave up farming and moved to Edinburgh, where he soon became one of a lively scientific circle, including the chemists Dr James Black and Sir James Hall, and John Playfair, Professor of Natural History at the University. Hutton travelled widely through Scotland, everywhere observing the rocks closely. From deductions made from his careful observations, Hutton developed his *Theory of the Earth*, which was published in 1788 in the first volume of the *Transactions of the Royal Society of Edinburgh*. A violent attack on this paper by Richard Kirwan, an Irish chemist and mineralogist, led Hutton to reply by supporting his original paper with *Proofs and Illustrations*. His slightly amended original paper and the supporting evidence were published in two volumes in 1795. Hutton, lucid and clear in conversation, became very involved and difficult to follow when pen in hand. Fortunately, his friend John Playfair presented Hutton's views with great clarity in his *Illustrations of the Huttonian Theory of the Earth* published in 1802. It was largely due to Playfair's attractive style that Hutton's views soon became widely known. Parts of Hutton's theory of the Earth are rather outside the scope of this book, in particular his view, based on much field evidence, that granite had been formed from molten rock material forced into pre-existing rocks and not, as was firmly held by Gottfried Werner (1749–1817), as a chemical precipitate in a universal ocean. Werner, who was a Professor of Mining at Freiburg in Saxony, was a famous teacher who attracted scientists from all over Europe. But Werner's theories, which he regarded as of world-wide application, were based on the study of the rocks of his native Saxony. These rival views were only resolved after years of controversy, often extremely bitter and outspoken, between the Plutonists (the followers of Hutton) and the Neptunists as the supporters of the Wernerian doctrine were named. As an example of the way in which the Huttonian principles came to be accepted, we may instance the eminent German geologist, Leopold von Buch (1774–1852). By training an ardent disciple of Werner, von Buch was converted to the Plutonist viewpoint as a result of his field studies of the volcanic areas of Italy, the Auvergne district of central France, the Canary Islands and the older rocks of Norway, Germany and the Alps.

We are primarily concerned here with Hutton's contributions to Earth-history. These were profound and revolutionary. Hutton interpreted the past by invoking only the processes that he could see acting slowly, often extremely slowly, in the present. He argued that wind and rain were gradually wearing down the 'everlasting' hills. The products of their destruction were transported to the seas in the form of river-borne mud and sand and there deposited to form new rock layers, which in their turn would be uplifted by forces within the Earth to form new lands and mountains. In an oft-quoted phrase he could see 'no vestige of a beginning – no prospect of an end'. By this he implied not that there was no beginning, but that as a result of so many changes, all vestiges of the beginning had been lost.

Hutton's dictum of 'the present as the key to the past' or the 'Doctrine of Uniformitarianism' as it is often termed, was vigorously attacked by the theologians, by the Neptunists, and by all those who preferred to invoke catastrophic events. The last embers of the controversy did not die down until well into the second half of the nineteenth century.

The general acceptance of Uniformitarianism with its implicit assumption of the long duration of Earth-history owes much to the work and writings of Sir Charles Lyell (1797–1875). Unlike Hutton, Lyell wrote persuasively and attractively. A Scottish gentleman, who had studied geology at Oxford under Dean Buckland's orthodox religious and catastrophic teaching, Lyell had private means which enabled him to travel widely through Europe. He soon became convinced that the features that he saw could be explained, not by great catastrophies, but by natural agencies acting over long periods of time. In 1830 appeared the first and in 1832 the second volume of his *Principles of Geology or the Modern Changes of the Earth and its Inhabitants*, in which he supported his arguments with a wealth of data. Lyell's *Principles* had a great impact on the scientific and well-read world of the time. This impact was maintained by the many later editions of the *Principles*, enriched by new evidence noted by Lyell on his travels to the United States of America and elsewhere, or communicated to him by friends, such as Charles Darwin. Lyell's other books and papers all stressed the Uniformitarian outlook.

Whilst the Doctrine of Uniformitarianism is one of the corner stones on which the science of geology is built, it has been argued that the 'present', that is the few centuries for which we have sufficiently detailed knowledge of the changes in or persistence of the form of the features of the Earth's surface, is far too short to be representative of all the events which may have occurred during a geological 'past' extending backwards, as we now know, for thousands of millions of years. This may well be true in certain respects. For instance, in the more ancient rocks of several continents there are vast deposits of iron-ores, which we cannot interpret in terms of the present, for nowhere on the Earth's surface are rocks of this type known to be in the process of formation today. But there is overwhelming geological evidence that for at least two thousand million years, the range of temperature on the Earth's surface has been within the limits now experienced, that is from slightly below freezing point to little above one hundred degrees fahrenheit. If this had not been the case, life could not have developed.

3. The Relative Geological Time-Scale

The Stratigraphical Table, lists in their order of deposition, the succession of rock layers which can be recognized on and immediately beneath the Earth's surface. This table began to be hammered out little more than a century and a half ago. Previous to this, several scientists, such as John Strachey (1671–1743) in Somerset and G. Christian Füchsel (1722–73) in Thuringia as well as Steno (p. 6), had shown that definite layers of strata could be recognized, forming either horizontal or inclined sheets, but their observations and conclusions were confined to small areas, and they had recognized their strata only by certain characteristics which gave them a distinctive appearance.

William Smith (1769–1839) or 'Strata' Smith as he came to be nicknamed, is regarded as the Father of Stratigraphical Geology. Unlike Hutton and Lyell, Smith had to support himself. Born of yeoman stock, orphaned at an early age, he became a largely self-taught surveyor and engineer. Employed in the construction of the Kennet-Avon Canal on the border of Wiltshire and Gloucestershire, he studied not only the rocks exposed in the cuttings, but also collected the abundant fossils which they contained. Collectors of fossils before him do not seem to have troubled to note the layers in which they occurred, but Smith was most careful to do this. He found that the different fossils which he could recognize did not occur indiscriminately, but that each was restricted to certain layers. Extending his work to the quarries of the neighbouring countryside, he was able not only to trace his succession of strata, but he found also that each layer contained its own particular and distinctive assemblage of fossils. Smith, working alone, had not only rediscovered Steno's Law of Superposition, but had made a most important addition – to use his own words 'the same strata were found always in the same order of superposition and contained the same peculiar fossils'. In 1779 he became acquainted with the Reverend B. Richardson, who had made a large collection of fossils from the area around Bath. Richardson was most surprised when Smith could tell him not only the locality but also the precise layer in which each fossil had been found. At Richardson's suggestion Smith drew up a table of the strata which he had recognized between the Chalk and the Coal giving their thickness, characteristic fossils and places where each occurred.

In his profession as an engineer Smith travelled widely, as much as 10 000 miles a year, an impressive figure for the days when only horse-drawn transport was available. Wherever he went, he observed most keenly the nature of the rocks and the form of hill and vale, steadily accumulating the data for his great geological map of England and Wales on the scale of 1 inch to 5 miles. This showed in colour the areas where each of his stratal divisions cut the Earth's surface, or outcropped. Smith's map was published in 1815, after several years of struggle to find both a publisher and also the necessary financial backing. In an explanatory text Smith set out his table of strata, using local terms such as Cornbrash, Portland Rock, London Clay, etc. Fig. 1, Smith's geological section from London to Snowdon, shows both his terminology and also his keen appreciation of the way in which certain strata formed either vales or scarps (lines of hills).

During the early part of the last century in different parts of Europe and in the eastern United States, other geologists were hammering out their own stratigraphical successions. Beds which yielded the same assemblage of fossils, were regarded as contemporaneous, even though they might be of rather differing lithology. As shown in Fig. 2, the tracing of the successive faunas enabled the beds exposed in different quarries or over wider areas to be connected together to

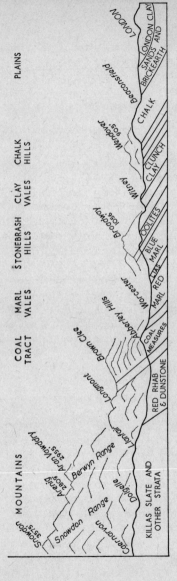

Fig. 1. *William Smith's section from Snowdon to London* (1817). *(Modified from the reproduction in T. Sheppard's William Smith: His Maps and Memoirs. Proc. York geol. Soc., 1917,* **19**, *Plate XIV.)*

build up a stratigraphical succession, composed of units or formations to which local names, of the type used by Smith, were given.

As knowledge of the successions in different areas in-

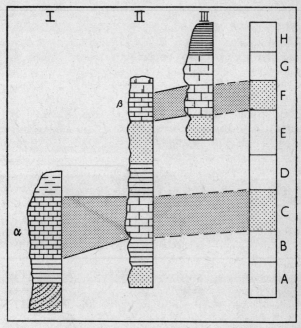

Fig. 2. *The Correlation of Strata. Limestone (brick ornament), clay (horizontal lines) and sands and sandstone (dotted) are exposed in the quarries* I, II *and* III, *but there are minor differences in lithology as indicated by the slight changes in ornament.* α *and* β *are two distinct faunal assemblages. By tracing these through the quarries, the beds can be correlated to build up the stratigraphical succession shown on the right.*

creased, it became apparent that there were major groupings of rocks or systems that could be traced through the plethora of local successions. These systems, each recognizable by its distinctive fossil assemblage, are units of international status, traceable throughout the world. But this universal Stratigraphical Table was developed before the days of international congresses and agreements and therefore the various systems were named somewhat illogically. Some from the areas where the particular system was first recognized, e.g. Cambrian, after the Roman name for Wales, Devonian, Permian after Perm now Molotov on the River Volga, Jurassic after the Jura Mountains. The Ordovician and Silurian Systems commemorate ancient British tribes who fought against the Romans in the borderlands of England and Wales. The Carboniferous and the Cretaceous (after the Latin word for chalk) Systems take their names from distinctive rock-types which make up a large part of each system. Triassic is from the threefold lithological succession of sandstone, limestone, claystone present in the type area of Germany. Palaeocene (*ancient recent*), Eocene (*dawn of the recent*), Oligocene (*little recent*), Miocene (*less recent*), Pliocene (*more recent*), Pleistocene (*almost recent*) and Holocene (*wholly recent*) are more logical, for their names, following a suggestion of Sir Charles Lyell, are intended to show that the older the system the more different is its fossil content from the forms of life now inhabiting the Earth.

The systems are grouped, following a suggestion of John Phillips, William Smith's nephew, into major units or eras, Cainozoic (*new life*), Mesozoic (*middle life*) and Palaeozoic (*old life*). The last two of these terms have largely replaced those used in an alternative method of naming the eras: Quaternary, Tertiary, Secondary and Primary.

American geologists prefer to divide the rocks of the Carboniferous System into two distinct systems, the Mississippian below and the Pennsylvanian above, but the majority of European geologists do not accept this.

There is also considerable difference of opinion as to the status of the units which have been recognized in the Tertiary strata; by some the divisions from the Palaeocene to the Holocene are regarded as separate systems, by others but as parts of one all-embracing Tertiary System, whilst a third view is to divide the rocks of Tertiary age between the Neogene and the Palaeogene Systems. Many French geologists prefer to name the Palaeogene rocks, the *Nummulitique*, after the nummulites, one of the most distinctive fossil groups to be found in these beds.

These problems of nomenclatural status will no doubt be resolved in due course by international agreement. They do not affect the order in which the different systems or quasi-systems were laid down.

The major units of the Stratigraphical Table can therefore be set out below:

Cainozoic	Quaternary	{ Holocene Pleistocene	
	Tertiary	{ Neogene	{ Pliocene Miocene
		{ Palaeogene	{ Oligocene Eocene Palaeocene
Mesozoic		{ Cretaceous Jurassic Triassic	
Palaeozoic	Upper	{ Permian Carboniferous Devonian	{ Pennsylvanian Mississippian
	Lower	{ Silurian Ordovician Cambrian	

These rocks from the Cambrian upwards contain fossils, in some places but sparsely, elsewhere in great abundance. For this reason they are often referred to as the Phanerozoic (*evident life*) in contrast to the rocks underlying the Cambrian strata. In these Pre-Cambrian or Cryptozoic (*hidden life*) beds, traces of fossils are not only exceedingly rare, but the organic origin of many of the so-called Pre-Cambrian 'fossils' has often been strongly disputed.

The rocks that comprise a system must have required time, indeed a very considerable amount of time, for their deposition. So the Stratigraphical Table is also a time-scale, giving the succession of periods, etc., but it is only a relative time-scale. The Cretaceous Period was obviously later than the Jurassic Period, for the rocks of the Cretaceous System overlie the rocks of the Jurassic System, but there is no means of telling from the table alone how much later in terms of absolute units, such as years or millions of years. The development of the absolute time-scale will be dealt with in the succeeding chapters.

Systems are divided into smaller units, usually on a tripartite basis into an Upper, Middle and Lower Series, though for certain systems, such as the Cretaceous, it is generally customary to recognize only an Upper and Lower Series. The name Epoch is given to the period of time during which a series was deposited.

Systems and Series are made up of Stages. This term dates from the work of the Frenchman, A. d'Orbigny (1802–57). He subdivided the Jurassic and Cretaceous rocks of France into a number of *étages*, each characterized by a distinctive faunal assemblage. Despite the fact that d'Orbigny held catastrophic views (p. 5), for he wrote 'I take for my starting-point . . . the annihilation of an assemblage of life-forms and its replacement by another', his basic concept of stages as major units for correlating beds over wide distances was correct. Although over a century of increasingly more refined study of the stratigraphical palaeontology of the richly fossiliferous rocks of the Jurassic system has passed, the major subdivisions now recognized in the Jurassic rocks of Western Europe are virtually the same as d'Orbigny's *étages*.

Age is the name given to the period of time during which the beds that form a stage were deposited. Stages are usually named by adding the suffix *-ian* to the type locality where the particular stage was first recognized, e.g. Ashgillian, Aptian, Portlandian, etc.

We are not concerned in this book with the minor and

more local units, called zones, which make up stages, or with the many types of zones that have been recognized and whose status and significance is distinctly controversial.

To summarize, there is a double terminology of stratigraphic units, represented by rocks, and of time units as given below:

Time Unit	Stratigraphic Unit
Era	—
Period	System
Epoch	Series
Age	Stage
—	Zone

BREAKS IN THE STRATIGRAPHICAL SUCCESSION

Where strata are all lying parallel, like a pile of books or magazines, they are said to be conformable. Hutton realized the significance of the unconformity exposed at Siccar Point on the Berwickshire Coast (Fig. 3). The lower succession of vertically inclined or steeply dipping beds must have been tilted and eroded, after which the higher group of almost flat-lying red sandstones was deposited across the uneven edges of the older rocks. He fully understood that the periods of formation of the two rock groups must have been separated by a long interval of time. Later work has shown that the red sandstones are of Upper Devonian, the vertical beds of Middle Silurian age. Therefore the tilting and erosion of the older series must have taken place during the Upper Silurian – Middle Devonian interval.

It is by no means exceptional to find two groups of rocks showing unconformable relationships. At Siccar Point the difference in dip is almost 90°, elsewhere it may be only a few degrees and, as we shall see in the next chapter, there are other and more subtle lines of evidence to show that the deposition of even a conformable succession of beds was not a completely continuous process.

In no one country can one find representatives of all the systems of the Stratigraphical Column. Everywhere the succession is liable to be broken by major unconformities, causing big gaps. In the British Isles, a greater range of systems crops out than in any other area of comparable size in the world, but even here, there is virtually no trace of the Upper Tertiary, for it is only in parts of East Anglia that beds of Miocene and of Pliocene age are to be found and these deposits are of very scanty thickness.

Fig. 3. *Sketch of the unconformity recognized by Hutton at Siccar Point, Berwickshire. The hammers lie parallel to the dip of the two groups of strata.*

THE MAJOR GROUPS OF ROCKS

So far, we have been considering only one of three major groups of rocks – the *sedimentary* rocks, such as clay, sandstone, limestone and coal. The sedimentary rocks are those which usually show bedding or stratification; those which often contain fossils, especially in certain layers; those which are made up, wholly or to varying extent, of the fragments of pre-existing formations, such as a conglomerate which is obviously a consolidated gravel or pebble bed, or organic remains, such as a coal seam or a reef-limestone, or chemical precipitates, such as a bed of rock-salt.

The second major group is the *igneous* rocks. These have been formed by the cooling and consolidation of molten material either on or below the Earth's surface. The igneous rocks are dominantly crystalline, being composed of tightly interlocking and usually sharp-edged crystals of a limited range of minerals. In the coarse-grained igneous rocks, e.g. granite, the individual crystals may be several inches in length, in the medium-grained rocks, such as dolerite, the individual crystals are just visible to the naked eye, whilst in the finer-grained rocks they can only be seen through a good hand lens or a petrological microscope. The fine-grained rocks grade down into glassy rocks, such as obsidian, which do not contain crystals. The texture of an igneous rock is broadly speaking an expression of its cooling history; the coarser textured rocks having cooled more slowly than those with tiny crystals.

Igneous rocks may occur in either *concordant* or *discordant* relationships with the sedimentary rocks (Fig. 4). The concordant igneous rocks are lava flows, which spread across a pre-existing surface, and sills which have been intruded along the bedding planes of sediments. The main discordant rock bodies are sheets or dykes, with thicknesses measurable in inches or feet, and batholiths which are irregularly shaped bodies of granitic rock, many square miles in area.

The last major group of rocks is the *metamorphic* rocks, which are formed when either sedimentary or igneous rocks have been affected by sufficiently high pressure or temperature or both for new minerals and textures to be developed. Clay rocks are commonly converted into slate, limestones into marble, and rocks of more mixed composition into foliated gneiss and schist, coloured by such typical metamorphic minerals as red garnet, colourless or black lustrous mica or greenish talc. Metamorphic rocks are also usually strongly contorted, clear evidence of the great pressures and stresses to which they have been subjected.

 OROGENIES OR MOUNTAIN BUILDING EPISODES

To west of the valley of the River Exe in Devonshire, gently eastward dipping sedimentary rocks of Permian age rest

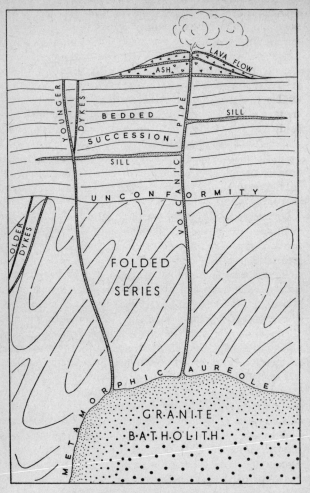

Fig. 4. *The relationship of different bodies of igneous rocks to the 'country rocks'. The size of stippling is related to the differences in grain size of the igneous rocks.*

unconformably on somewhat metamorphosed rocks folded along axes trending east-to-west. The metamorphism has not been intense enough to destroy all trace of fossils, so we know that these folded rocks were originally sediments of Devonian and Carboniferous, though not latest Carboniferous, age. Into the folded rocks were intruded the granites of Land's End, Dartmoor, etc. Both the granites and the metamorphosed country rocks are cut by numerous dykes and mineral veins. As pebbles of dyke rocks, mineral veins and the marginal rock-types of the batholiths are to be found in the Permian sediments, the folding and the intrusion of the igneous rocks must have occurred before the Permian rocks were deposited.

Brittany, parts of the Massif Central of France, the Vosges Mountains, the Ardennes, the Rhenish Schiefergebirge, the Black Forest and the Harz Mountains, are all areas largely or in part composed of Upper Palaeozoic rocks and are surrounded by unconformable sediments of Permian and later Mesozoic age. The upper Palaeozoic rocks have been strongly folded, indeed in some areas they have been considerably metamorphosed and have been intruded by granites and other igneous rocks. These areas are the worn-down stumps of a former mountain chain, the Variscan Chain, which was produced when the sedimentary rocks of these areas were affected by orogenic or mountain building movements in late Upper Palaeozoic times.

Another belt of strongly folded and altered rocks extends down the west side of Scandinavia through Scotland into the Lake District and Wales. But this mountain chain, the Caledonides, is considerably older than the Variscan belt, for the folded sediments of Lower Palaeozoic age are overlain unconformably by beds belonging to the Devonian System. Yet another belt of folded rocks extends across southern Europe from the Pyrenees, through the Alps to the Carpathians and then eastwards across Asia to the Himalayas. These Young Mountain Chains have a much greater relief and grandure than either the Variscan belt or the Caledonides, for geologically speaking they have only recently

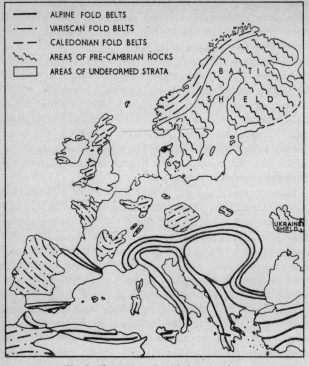

Fig. 5. *The main structural elements of Europe.*

been uplifted and exposed to the attack of the agents of erosion. The main movements of this latest mountain building episode, the Alpine orogeny, occurred in mid-Tertiary times.

As shown on Fig. 5, we can therefore trace in the Phanerozoic rocks of Europe three mountain chains, whose periods of formation were separated by vast periods of time. The outcrops of the folded and contorted rocks are ringed round by great spreads of almost flat-lying and unaltered sediments that were formed in part by the erosion of the earlier-formed mountain chains.

As we have seen, the pioneers of Stratigraphical geology were mainly Europeans and it used to be thought that these three orogenic episodes, the Caledonian, the Variscan and the Alpine, were major events in Earth-history of world-wide extent. As the rocks of other continents were studied, the same pattern of narrow elongated belts of strongly folded rocks separated by great stretches of flat or nearly flat-lying sediments was found. But it became clear that the orogenic episodes had not occurred at the same times as those in Europe. For instance, the rocks of the greatly degraded old mountain chains traceable for 2000 miles from Newfoundland, through New England and then south-westwards towards Florida, were folded during three orogenic episodes, the Taconic in the later part of the Ordovician Period, the Acadian in late Devonian and the Appalachian in Permian times. The Nevadan Orogeny, which profoundly affected the western side of the North American continent, occurred in late Jurassic – early Cretaceous times and was followed, towards the close of the Mesozoic era, by another phase of uplift, the Laramide Revolution.

It is now recognized that each orogenic episode was limited in its geographical extent. Recent work, however, has shown that each orogenic episode was spread over a considerably greater time-span than was formerly thought to be the case. Hutton's unconformity at Siccar Point (p. 17), with Devonian beds resting with strong unconformity on Silurian strata is one result of the Caledonian folding. It used to be thought that the main Caledonian folding occurred during this post-Silurian – pre-Devonian interval, but the more thorough investigation of the folded Lower Palaeozoic rocks of Wales, the Lake District and the southern Uplands of Scotland has brought to light comparable unconformities at many other stratigraphical levels. The strongly deformed and metamorphosed rocks of the Highlands of Scotland are overlain unconformably by flat-lying beds of Devonian age. Techniques developed in the last twenty years for analysing the deformation to which metamorphosed rocks have been subjected have shown that the rocks of the Highlands have suffered at least three distinct episodes of folding. Early folds have been refolded during later periods of movement as shown in Fig. 6, often along axes of different trend. In some areas only, the effects of the last deformation can be seen, for it has been so intense that it has overprinted (obliterated) the effects of earlier episodes. In other areas, however, the effects of earlier movements can be disentangled and a fold-history elucidated from the scattered pieces of evidence, in much the same way as the Stratigraphical Table was first constructed. But in areas which have been so strongly metamorphosed that all trace of

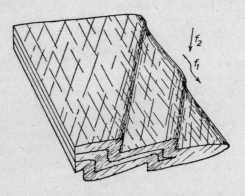

Fig. 6.   *Refolded fold. (From J. G. Ramsay, Fig. 11B, p. 160 in* The British Caledonides, *Oliver and Boyd, 1963.)*

fossils have been destroyed, we can, at present, only build up a sequence of movement episodes, M_1, M_2, (see p. 102) etc., without being able to relate them directly to the Stratigraphical Table and say definitely that M_1 is of the Lower Ordovician or of Upper Ordovician age. In the Highlands of Scotland a distinction is made between the Older and the Newer Granites. The margins of the Newer Granites cut across the structures of the folded country rocks. These Granites must be post-tectonic, for they were clearly intruded after the climax of the folding; on the other hand the Older Granites are not only of slightly different mineralogical composition, but their margins are much less clean-cut, for they tend to grade into the country rocks through a marginal zone of granitoid gneiss. Such granites are syntectonic, for they must have been emplaced during the folding.

The Caledonian movements in Britain are now recognized as having begun in post-Upper Cambrian – pre-Lower Ordovician times. Indeed some would put their commencement even earlier. Intermittent movement, varying in intensity and in locality, continued throughout the Ordovician and Silurian Periods until post-Lower – pre-Upper Devonian times. The succeeding Variscan movements were equally long drawn out. Beginning in very early Carboniferous times in the south of the Variscan belt (see Fig. 5), the movements gradually spread northwards and did not seriously affect the south of the British Isles until towards the close of the Carboniferous Period. In the same way, whilst the climax of the Alpine folding affected the western Alps in post-Oligocene – pre-Miocene times, there were important late Cretaceous movements in the eastern Alps and, in the Pyrenees, there were both Upper Cretaceous and post-Eocene phases. The final uplift of the Alps and the Pyrenees did not occur until the Pliocene.

EPEIROGENESIS OR VERTICAL MOVEMENTS ON A REGIONAL
SCALE

That the rocks of the orogenic belts have suffered strong lateral compression is shown both by their metamorphism and also by the presence of nearly horizontal thrust planes along which great masses of rock have been displaced for distances measurable in miles. The unmetamorphosed rocks of the intervening areas, the basins and the platforms, have also suffered some disturbance, but of a much more gentle and subtle nature. Careful tracing of individual horizons often shows that a particular bed may not always be underlain by the same stratum but that it may overstep across on to older rocks. The difference in dip between the two rock groups may be only a few degrees; indeed in any one exposure the two groups may appear to be conformable. It must be remembered that a difference in dip of 1° between adjacent strata, is not usually recognizable in an exposure, for bedding planes are not plane surfaces, they are nearly always slightly uneven and irregular. If the difference in dip of 1° between two bedding planes at a particular point persists, then one mile away the same bedding planes will be separated by nearly 100 feet of strata.

Modern and extremely detailed stratigraphical studies have shown that the almost flat-lying rocks of the basins and the platforms have been affected by intermittent but persistent epeirogenic movements, demonstrable by the tracing of oversteps, like that shown in Fig. 7. In this area, the beds older than the Upper Greensand must have suffered slight eastward tilting, accompanied locally by gentle folding and slight faulting, before they were submerged beneath the sea in which the Upper Greensand was deposited. Epeirogenic movements are therefore essentially vertical, causing the upwarping or downwarping of areas of considerable extent.

THE PRE-CAMBRIAN

So far we have only been concerned with the rocks of Phanerozoic age, but these rest on great thicknesses of the Pre-Cambrian. In the British Isles, apart from the Highlands of Scotland and parts of Donegal and Connemara in Ireland, the outcrops of the Pre-Cambrian rocks are of very limited extent. It is only in a few widely scattered areas, such as parts of Anglesey, the Longmynd in Shropshire and the Malvern Hills, that the Pre-Cambrian rocks project through the cover of Phanerozoic beds. But in other parts of the world, Pre-Cambrian rocks occupy great areas, thousands of square miles in extent. These Shields, as they are called, such as the Baltic Shield, the Canadian Shield, the African Shield, etc., are regarded by some as the nuclei of the continental masses.

Till little more than a decade or so ago, the Pre-Cambrian rocks were usually subdivided into Younger Pre-Cambrian,

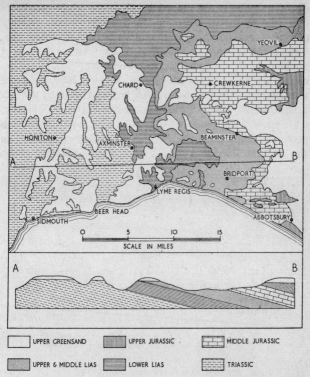

Fig. 7. *Map and section showing the westward overstep of the Upper Greensand across the Jurassic and the Triassic rocks in west Dorset and east Devon.*

the so-called Algonkian or Proterozoic (*earlier life*), composed of rocks which were virtually unmetamorphosed or but slightly metamorphosed, and an Older Pre-Cambrian, the Archaean or Archaeozoic (*primaeval life*) of intensely metamorphosed beds. Recent work has shown that the story is much more complex.

For instance, the spectacular scenery of Wester Ross along the north-west coast of Scotland owes much to a line of isolated mountains such as Canisp, Suilven and Quinag.

These mountains are capped by gleaming white quartzites
(fossil beach sands), overlain conformably by beds yielding
marine fossils of unquestionable Lower Cambrian age.
These quartzites rest unconformably, but with a difference of
dip of only a few degrees, on reddish sandstones and shales,
named the Torridonian after Loch Torridon. The surface of
unconformity is remarkably even and clearly is an erosion
surface across which spread the waters of the early Cambrian
sea. Around the flanks of the mountains, another erosion
surface is easily traceable, but this is markedly irregular (see
Fig. 8). The Torridonian rests on strongly contorted gneiss

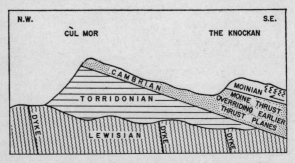

Fig. 8. *Sketch section showing the relationship of the major
rock groups near Loch Assynt, Sutherlandshire.*

and schists, named the Lewisian Complex, for this older
group forms Lewis and the other isles of the Outer Hebrides.
The characters of the Torridonian show that it was laid
down on land, burying a land surface of quite considerable
relief that had been cut across the Lewisian rocks. Early
workers in the area had suggested that the Lewisian rocks
must have had a long and complex history, but it was not
until 1950 that this history began to be unravelled. Then it
was shown by J. Sutton and his wife Janet Watson, that the
rocks of the Lewisian Complex had been affected by two
major orogenic episodes; an earlier one, named the Scourian
and a later one, the Laxfordian. In some areas the effects of
the Laxfordian orogeny were so intense that they have over-

printed and obliterated all trace of the Scourian metamorphism, but other areas were less affected by the later earth movements and there the effects of the two orogenies could be disentangled. The clue was given by dolerite dykes, which were intruded after the Scourian metamorphism and were in their turn affected by the Laxfordian Orogeny. In the areas that had not been overprinted by the Laxfordian Orogeny, the dykes are steeply inclined parallel sheets, but as they are traced into the Laxfordian areas, the dykes not only are recrystallized, the dolerite being altered to amphibolite, but they also become complexly folded.

It was suggested from purely geological evidence that there was a very considerable time gap between the Scourian and Laxfordian episodes, for the dykes must have been intruded into thoroughly cold and brittle rocks. In a later chapter we shall see how the development of isotopic dating has enabled these events to be dated in terms of millions of years, whilst the latest work indicates that the story of the 'Lewisian' is even more complicated than the outline given above.

It has long been known that the Pre-Cambrian history of the shield areas is both long and complicated. The so-called Archaean rocks represent the worndown stumps of a number of mountain chains, but in such strongly metamorphosed rocks, it is often very difficult to establish a relative chronology of these orogenies. Further, even if a relative chronology can be established in, say, the Canadian Shield, it is impossible by normal geological methods to establish a satisfactory correlation with the orogenic sequence of the Baltic Shield. There is no fossil evidence, as there is in the Phanerozoic rocks, to provide the key for intercontinental correlation. But as we shall see in the last two chapters, the position has been drastically changed in the development of isotopic dating. Not only is intercontinental correlation of the episodes of Pre-Cambrian time becoming possible, though admittedly as yet only on a very broad scale, but also it has been shown that certain areas of so-called Archaean rocks do not in fact belong to the older Pre-Cambrian, but are rocks which have been intensely metamorphosed by localized orogenies in late Pre-Cambrian times.

We have been concerned in this chapter with the relative time-scale that can be erected by what may be called orthodox geological methods. By these the succession of the beds and of the orogenic and epeirogenic movements which have affected them can be worked out in detail. Indeed, in those areas and in those parts of the Stratigraphical Column which yield diagnostic fossils in sufficient abundance, the sequence of events can be worked out in very considerable detail. This applies to the beds ranging in age from the Holocene to the early Cambrian. But below the base of the Phanerozoic, the picture becomes blurred. Locally, as in Wester Ross, we can establish a relative sequence of events for the Pre-Cambrian strata that are exposed. In the Shield areas it is clear that the Pre-Cambrian rocks must represent a very long period of time. But by geological methods alone, it is impossible to extend a Stratigraphical Table for world-wide use down into the Pre-Cambrian, nor can one compare, with any degree of precision, the duration of Cryptozoic or Pre-Cambrian times with that represented by the Phanerozoic rocks of the Stratigraphical Table.

4. Primarily Geological Estimates of a Time-Scale

The sedimentary rocks have been formed by the breakdown of pre-existing rocks. By the action of the wind and ice, the expansion and contraction caused by temperature changes, and the attack of very dilute solutions of solvents, in rain-water, acting over a sufficiently long period of time, rocks, are either broken down into fragments of smaller size or their more soluble constituents may be removed in solution. The material thus formed, whether in the solid state or in solution, is transported by moving water, by glaciers and ice-sheets, by the wind or by mass movements under gravity, to the depressions on the Earth's surface. Such depressions are ocean basins and shallow seas and also lakes and other lowlying areas on the land surface.

New sedimentary rocks are formed in these depressions. The clastic sediments, such as conglomerates (cemented pebble beds) or sandstones are clearly composed of rock fragments or of the more resistant minerals derived from pre-existing rocks. The pyroclasts, including the coarse-grained agglomerates and the finer-grained ashes, are made up of material which has been blown out into the atmosphere by volcanic eruptions and has then settled either on land or in water. Other types of sediments, such as the coal seams or certain limestones are primarily of organic origin, for they are composed either of closely packed plant fragments and plant debris or of the remains of organisms, such as corals and other reef-builders, which build up their skeletons from the calcium carbonate content of sea-water. Other types of limestone are termed bioclastic, for they are

composed mainly of organic fragments, which have been
rolled and abraded as they were transported by waves or by
currents from their place of life to the place of deposition.
Yet other sediments, such as the beds of rock-salt, anhydrite,
gypsum or potash-rich salts, form the evaporite deposits and
were deposited as chemical precipitates in shallow seas or
lagoons. The fine-grained calcitic matrix in which the organic
fragments of many bioclastic limestones are set was precipitated from water super-saturated with calcium carbonate.
The clay rocks are mainly made up of extremely minute
aggregates of complex silicates, the clay minerals, produced
by the chemical weathering of pre-existing rocks and then
transported by moving waters into quiet regions, where they
have settled to the bottom.

Sediments are liable to be altered by water circulating
through their pore spaces either soon or long after deposition. Material may be precipitated to cement the quartz
grains of a sand into a sandstone or, if silica is the cementing
agent, into a quartzite. In muds, chemical reactions may
occur aided by the action of bacteria. Sulphides may be
formed to react with carbon dioxide, forming carbonates,
which are precipitated in the pore spaces of the mud, so
altering its chemical composition. The flints of the Chalk
were formed by waters moving through the chalky ooze; in
certain layers these waters were able to dissolve the spicules
of sponges and the remains of other organisms which
secreted a silicous skeleton, while in other layers, with slightly
different pH, the silica was precipitated to replace the extremely fine-grained limestone and form nodules or bands of
black flint. Post-depositional diagenetic changes also include
those due to the weight of superincumbent sediments,
squeezing out the pore water, pressing the particles into
closer contact, so that the sediment becomes compacted and
lithified.

The thickness of sediment that can accumulate is determined not merely by the amount of detritus supplied, but
also by the rate of subsidence of the basin into which that
material has been moved. If the rate of subsidence is equal to

or greater than the rate of supply, then steady accumulation can occur, but if the rate of supply overtakes the rate of subsidence, then sooner or later, a point will be reached when deposition will not be able to continue uninterruptedly, for waves and bottom currents will begin to work over the material. At first all material below a certain grain-size will be swept away and only the coarser material will be left, producing a condensed deposit. If subsidence is halted, or perhaps even reversed by slight vertical uplift, then submarine erosion may take place with the formation of a *remanié* pebble bed, made up of rolled and abraded fragments of the hardest parts of the previously deposited sediment, or under more extreme conditions, of a 'hard ground' of bare rock.

Recent investigations of the nature of the shelf seas bordering the coasts of Europe and the United States have provided abundant evidence of the extremely patchy and irregular nature of modern sedimentation. There is abundant evidence that similar conditions prevailed in the shallow shelf seas in which Jurassic and Cretaceous rocks of Europe were deposited. Oversteps at many horizons of the kind shown in Fig. 7 show that epeirogenic movements were causing restlessness of the floor on which these sediments were accumulating. At numerous levels in the limestones of the Cotswolds and also in the Chalk, one finds erosion surfaces or fossil hard grounds. In the oolitic limestones of the Cotswolds the upper surface of a bed of limestone will be planed off, riddled with borings and often encrusted with oysters, whilst overlying it will be a *remanié* pebble bed (Fig. 9). In the softer Chalk, one finds at intervals thin beds of nodular chalk made up of rolled pebbles of chalk with their upper surface bored, whilst the bed is speckled with grains of greenish or black glauconite, a distinctive mineral known to be forming today off the California coasts, on parts of the sea floor where there is little or no sedimentation. Another feature of both recent and fossil examples of pauses in deposition is the presence of pebbles and shells rich in calcium phosphate. Even the thick marine clays of Jurassic and

Cretaceous age were not everywhere deposited without pause; this is shown by the presence of seams of phosphatic nodules and also by thin beds in which abraded and broken shells are concentrated. Similar evidence of pauses in deposition is to be found in the marine limestones and clays of all other systems.

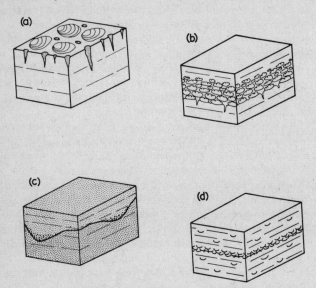

Fig. 9. *Evidence of pauses in sedimentation.*
(a) *A bored and oyster-encrusted erosion surface in limestone.*
(b) *A bored surface overlain by rolled and bored pebbles of limestone.*
(c) *A bed of sand channelling down into the underlaying layer.*
(d) *Thin layer of abraded shells in a succession of clays.*

Interruption in deposition may also occur on land. When the infill of a fault-bounded intermontane basin reaches a certain level, its upper surface is liable to be worked over by the wind and by intermittent sheet floods. Lake basins eventually become silted up.

The folded mountain chains (p. 21) mark the site of former geosynclines. These are elongated belts of the Earth's crust, which have steadily subsided throughout long periods of time. Into the geosynclinal troughs poured the waste from the bounding areas of land and from islands. The sandstones and grits of the geosynclinal (greywacke) facies are characteristically poorly sorted, being composed of angular to sub-rounded grains of quartz and feldspar, and rock fragments set in a finer grained-groundmass. They contrast with the clean-washed, well-rounded quartzose sandstones typical of shelf seas. Limestones, in general, form but a small proportion of the total thickness of sediments and when they do occur, as in parts of the Tethyan geosyncline from which the Alps were formed, many of the limestones show evidence of slow and interrupted deposition. Such 'starved' deposits must have been formed in the central parts of the troughs away from the rapidly subsiding marginal belts, which acted as traps for the more sandy and clayey sediments.

The Appalachians are the area in which the concept of geosynclines was developed. In 1859 James Hall showed that the mountain ranges of the northern Appalachians were formed by a thickness of 40000 feet of Palaeozoic sediments, showing throughout evidence of having been deposited in shallow seas. He further demonstrated that when these beds were traced westwards into the continental interior they decreased in thickness to a mere few thousand feet of strata. Hall thought that the steady depression of the trough was due to the weight of the sediments which infilled it. But in 1873 J. D. Dana, who suggested the term geosynclinal, later changed to geosyncline, argued that such a thickness of sediments could only be built up where steady subsidence of the crust allowed space in which they could accumulate.

The detailed study in recent years of other geosynclines, has shown that whilst they differ in many points of detail from the original 'type' Appalachian geosyncline, they have suffered the same history. Eventually subsidence ceased and then the vast pile of sediments, often with interbedded volcanic rocks, was folded and compressed, intruded by

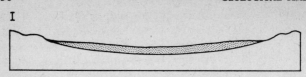

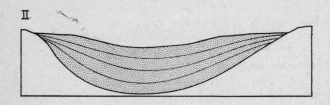

Fig. 10. *The history of a geosyncline*

   I. *First development.*
  II. *Maximum downsagging. This is often accompanied by the intrusion and extrusion of basic igneous rocks.*
 III. *The Orogenic Phase. The sediments are metamorphosed and intruded by granites (crosses). The granite body on the left is syntectonic, but that on the right cuts across the folds and is, therefore, posttectonic.*
 IV. *Uplift followed by erosion has produced a mountain chain.*

great bodies of granite and finally uplifted to form a mountain chain (Fig. 10).

The maximum thickness of the sediments formed during a particular period of geological time is therefore most probably to be found in the contemporary geosynclines. As we have argued above, sedimentation on the land areas or on the shelf seas of the continental margins was extremely unlikely to be continuous for long periods, whilst on the deeper parts of the ocean floors, deposits of organic ooze or of the very finest dust accumulate but extremely slowly.

There is, however, the practical difficulty of measuring the true thickness of beds that have been strongly folded. This can only be done, if the beds contain a number of thin distinctive horizons, so that by following these the complexity of the fold pattern can be unravelled. In a number of folded areas, the detection and tracing of such marker horizons has shown that rock units which were formerly thought to be of great thickness, for the strata all dipped uniformly in the same direction and at high angles over considerable distances, were in fact built up of tightly packed folds with the marker horizons outcropping repeatedly on either side of the fold axes. The thickness of the beds involved was measurable in scores rather than in the thousands of feet estimated before their fold style had been unravelled.

ESTIMATES OF THE THICKNESS OF STRATA

In 1878 the Irish geologist Samuel Haughton suggested that 'the proper relative measure of the geological periods is the maximum thickness of strata formed during these periods'. By adding together the maximum thicknesses of each system then known in any part of the world, he reached an estimate of 177200 feet for the total of the Phanerozoic rocks. A number of other geologists have made similar calculations, the figures ranging from 100000 feet by Geikie in 1892 to 452000 feet by Holmes in 1965. In general, the more recent the estimate, the higher the figure, but this largely reflects the growth of our knowledge of the geological make-up of more and more of the land areas. The Miocene System is represented, if at all, in the British Isles, by a *remanié* pebble bed little more than a foot in thickness present only in a small area in East Anglia, yet 24000 feet of Miocene beds have been recorded in California; the total maximum thickness of each of the Jurassic and Cretaceous Systems in England is less than 5000 feet, as contrasted with the over 45000 feet of Jurassic beds and nearly 50000 feet of Cretaceous rocks claimed from California.

On Fig. 11 the Stratigraphical Column has been drawn to a scale of maximum known thicknesses to comply with Haughton's suggestion for determining the relative duration of the different periods.

But, as we have argued in the preceding pages, Thickness of Beds = Rate of Subsidence — Effects of Contemporaneous Erosion, and even in geosynclinal deposits, there is unmistakable geological evidence that sedimentation was not entirely uninterrupted. A time-scale based on the thickness of beds must therefore contain many uncertainties and can, at best, be regarded as but an approximation to the truth.

Still more uncertain is the next step. If the average maximum rate of accumulation of the sediments is known, then the maximum thickness table can easily be converted into a time-scale. But we have at present accurate information as to the rate of sedimentation only from a very limited range of environments and these are mainly close inshore. It is possible with the modern core sampling tubes to bring up samples, often many feet in length, of the soft uncompacted sediments accumulating on lake floors or on parts, even the deeper parts, of the sea floor, but too often we do not know the number of years that a sample took to accumulate. Such samples have not been taken in the geosynclines of the present, such as the sea floors off the East Indian chain of restless islands. Estimates of the average rate of sedimentation that have been made show even greater variation than those given for maximum thickness, for they range from one foot in a hundred years (Sollas, 1909) to one foot in 8616 years (Haughton, 1871). Also it has to be assumed that the average maximum rate of sedimentation has remained the same from Cambrian times to the present, an assumption that may well be stretching the Doctrine of Uniformitarianism too far (see p. 8).

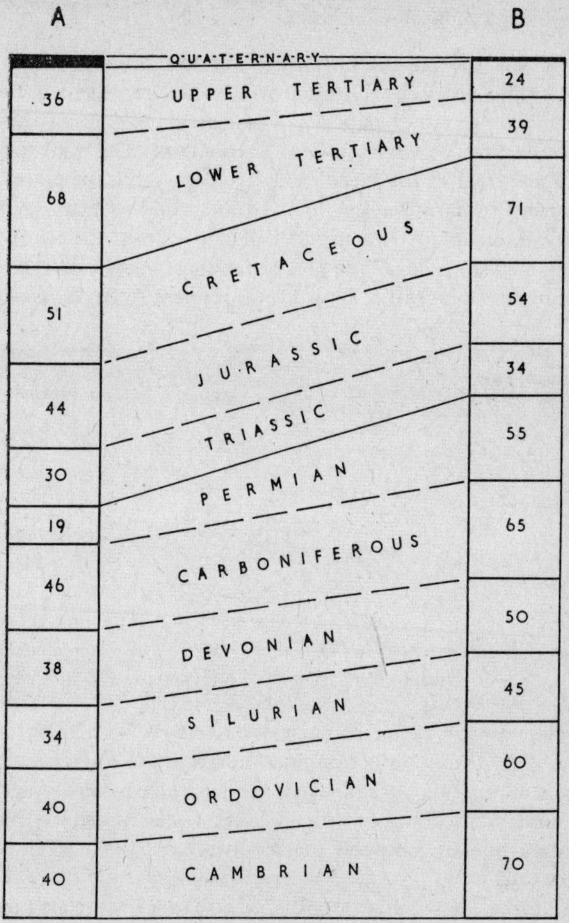

Fig. 11. *The Phanerozoic Scale.*
A. *On a basis of the maximum thicknesses in thousands of feet of the various systems (after Holmes 1959).*
B. *The isotopic timescale in millions of years. (After* The Phanerozoic Time-Scale, Geological Society Symposium, 1964*).*

CHRONOLOGY BASED ON VARVE ANALYSIS

In a few special cases, however, it has been possible to
determine precisely the number of years represented by a
particular group of deposits. The classic example of this is
provided by the varve clays of Scandinavia. The final stages
of the retreat of the Pleistocene ice-sheets from the plains of
North Germany back into the mountains of Scandinavia
were interrupted by a number of halts, when the ice-front
remained stationary. The position of these halts is marked by
a number of moraines (Fig. 12), prominent ridges formed by

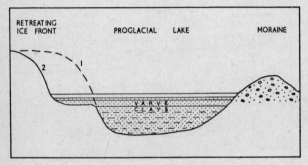

Fig. 12. *The formation of varve clays.*

the deposition at the ice-front of its load of sand, gravel and
boulders. Comparable terminal moraines are to be found at
the snouts of modern glaciers. When the ice retreated again,
a large number of proglacial lakes were formed by meltwaters ponded between the ice-front and the last formed
moraine.

During the summer months, when the ice is melting most
rapidly, the meltwaters will be whitish and opaque, the so-
called 'glacier milk', for they are highly charged with sand,
silt and clay from the material which is carried in and under
the glacier. This material, suspended in the water, gradually
settles to the bottom, the coarser grains falling most rapidly.
The finest material may not settle until the winter comes and

the lake freezes over. It forms a layer, darker in colour and of much finer grain than the material deposited earlier in the year.

The proglacial lake basins therefore become filled with a laminated succession of 'varve clays'. Each pair of laminae with a lower silty and sandy layer and an upper, more clayey layer represent the deposits laid down during a particular year. The laminae are normally only a few millimetres to a few centimetres in thickness, though occasionally in the succession there will be layers which are either abnormally thin or abnormally thick, perhaps up to some tens of centimetres. As the ice retreated, the last varves to be deposited will overstep the earlier varves in the direction of the ice retreat (Fig. 12).

The significance of such varve clays was appreciated by the Swede G. De Geer (1858–1943), who carefully measured in a large number of sections the thickness of each annual layer. Not only did this show the exact number of years represented by each section of varves, but he found that by comparing the records of adjacent sections, he was able to recognize 'marker horizons', abnormally thick or abnormally thin varves, or perhaps a distinctive combination of thick-thin-thick. He was therefore able to correlate his sections and so build up a succession, using the same principle as in Fig. 2. More important, through being able to relate his varve succession to the terminal moraines, he was able to count the numbers of years that had elapsed as the ice-front retreated from the position marked by one moraine to the next more northerly moraine. The rate of retreat of the ice-front proved to be rapid, averaging some hundreds of metres a year.

A significant point in the retreat of the ice-sheets was the so-called Bipartition, when the melting ice-sheets split into two portions. The last remnants of the southern position linger on in the ice fields of the Hardangerjökul near Finse, in southern Norway, whilst the remains of the northern portion are even more extensive. At Ragunda, near the place of Bipartition, there was a lake which had been accidentally

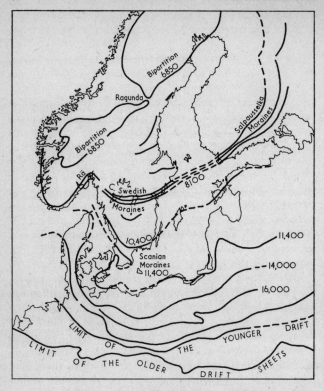

Fig. 13. *Map of the chief moraines laid down as the ice
sheets retreated into the mountains of Scandinavia. The
figures date the approximate positions of the successive ice
fronts in years* B.C. *(after various authorities).*

drained in A.D. 1796. The varves of this lake represented 3000
years and amongst them was a particularly thick varve,
which De Geer was able to recognize in many other sections.
He regarded this thick varve as caused by the great run-off of
ponded water due to the Bipartition. Unfortunately at Lake
Ragunda varve formation had ceased long before A.D.
1796, so it was not possible to date precisely the top of the
varve sequence. But another worker found a sequence of

varves in central Sweden, extending to historic times, and so it was possible to date precisely the top of the varve succession.

The major events of the last phases of the Pleistocene glaciation can therefore be dated precisely, as shown in Fig. 13, with less than 10000 years covering the retreat of the ice-front from north Germany to its Bipartition in the Scandinavian mountains. This last event occurred, according to De Geer and his school, about 6850 B.C. As we shall see later (p. 89), these dates agree closely with those obtained by radiocarbon dating, an entirely different technique.

The Scandinavian varve sequence is unique in its completeness. In North America, E. Antevs, one of De Geer's students, has attempted to establish a chronology for the retreat of the ice from Long Island near New York to northern Ontario, but unfortunately, there are in the sequence several large gaps whose magnitude has yet to be estimated, nor is it possible to establish a direct link with modern chronology. The estimate that the Connecticut moraines were deposited about 30000 years ago is therefore debatable.

OTHER EXAMPLES OF LAMINATED SEDIMENTS

Finely laminated sediments of a varve like character overlying unmistakable glacial deposits have been found at a number of localities in the beds laid down by the older (pre-Pleistocene) ice-sheets. This is especially the case in the deposits of the Permo-Carboniferous glaciation of the Southern Hemisphere. Similar sediments have even been found in widely separated areas in beds of younger Pre-Cambrian age. But if such sequences of varves are counted, they only give the time-span represented by these particular layers; they cannot be connected so as to provide a time-scale of more general application.

Laminated beds, with laminae which may be annual, have been claimed from a number of other horizons. In 1929, W. H. Bradley described the Green River Beds of Wyoming,

Colorado and Utah. These include a 2000-foot thick succession of lake deposits, showing a regular alternation of
laminae with one member of each pair darker and much
richer in organic matter than the other member. Bradley
interpreted this as an annual rhythm. By measuring and
counting these varves, which are not, however, glacial in
origin, he estimated the Green River Beds to have been
deposited during the passage of between five to eight million
years. His actual figure was six and a half million years with
an error of plus or minus one and a half million years. The
Green River Beds are of Middle Eocene age and if it is
assumed that lake deposits in the beds of Eocene age overlying and underlying the Green River Beds accumulated at
the same rate, then the estimated duration of the Eocene
Period is 22 million years with an error of plus or minus five
million years. This agrees reasonably well with the 16 million
years of the latest estimate using isotopic data (p. 85).

The Upper Permian evaporite deposits of Texas include
beds which show a closely spaced lamination of calcite-rich
and anhydrite-rich layers. Some geologists regard these as
annual layers, but others believe that they reflect oft-repeated
breaching of the barrier which separated the lagoons in
which these beds were deposited from the open sea. Immediately after a breach the shallow lagoons would be flooded
with water of normal salinity, but as soon as the breach was
closed, under the conditions of high temperatures and
evaporation rates which then prevailed, the salinity of the
water in the lagoon would be steadily increased so that calcium sulphate (anhydrite) would be precipitated instead of
calcium carbonate. Then at the next inflow of normal sea-
water, the precipitation of calcite would be resumed. If this
is true, then there is no reason for the layering being annual.

RHYTHMIC DEPOSITION

In parts of the Stratigraphical Column, a considerable
thickness of beds may be made up of a limited range of
lithologies, repeated time and time again, in the same

sequence, *A, B, C, D, A, B, C, D,* etc. Such cyclothems, as these rhythmic units are called, are particularly characteristic of the Carboniferous System, but the type of rhythm is not everywhere the same. In the parts of the Coal Measures of England and Wales, the unit is,

6 Coal seam
5 Seat earth or fossil soil (fire clay or ganister)
4 Sandstone
3 Shale with freshwater fossils
2 Shale with brackish water fossils
1 Marine band
   Coal Seam.

In the Yoredale Beds of late Lower Carboniferous age in the Pennines the rhythm is,

   Limestone with marine fossils
4 Thin coal seam (often absent)
3 Sandstone with drift plant remains
2 Shale
1 Marine limestone.

Whilst in the Pennsylvanian (Upper Carboniferous) beds of Illinois, a typical cyclothem shows,

8 Shale with concretions of ironstone
7 Marine limestone often with shaley partings
6 Shale with marine fossils
5 Coal seam
4 Seat earth
3 Unfossiliferous limestone
2 Shale with non-marine fossils
1 Sandstone, often with an uneven base channelling down into underlying shale.

The underlying controls of such rhythmic successions are a subject of vigorous debate. Clearly this alternation of widely traceable marine and non-marine horizons must be due to slight changes in the height of sea level relative to the

land areas. However this was caused, we have no evidence
at present to indicate that there was in any way a time
control, so these rhythmic units cannot be used, as are the
varve clays of Scandinavia, for time-measurement. The
same applies to the larger rhythmic successions which have
been recognized in the Permian evaporites of north-west
Europe and in the deposits of several systems which were
laid down in a deltaic or semi-deltaic environment.

THE SALT CONTENT OF THE OCEANS

In 1715 the Astronomer Royal, Edmund Halley, suggested in
a paper in the *Philosophical Transactions of the Royal Society*
that the age of the Earth could be determined by measuring,
at intervals of a few decades, the salt content of the seas, and
of lakes such as the Dead Sea that have no outflow. He
lamented that the Greeks and Romans had not made such
determinations for 'it cannot be doubted but that the
difference between what is now found and what then was,
would become very sensible'.

In 1899 the Irish geologist J. Joly estimated the age of the
Earth by calculating the amount of sodium carried to the
oceans by the rivers of the world and then dividing this
figure into the sodium content (10·56 parts per thousand) of
sea-water. He made a number of assumptions, first that the
waters of the oceans were originally fresh, secondly that not
only has the sodium content of river water remained constant, but also that the volume of river waters and the rate at
which sodium was removed from the rocks have all remained
virtually unchanged throughout geological time. This is
surely stretching the Doctrine of Uniformitarianism to the
limit.

But it may well be that the Earth's surface now shows an
abnormally great amount of relief. In many parts of the
world there is evidence of considerable orogenic activity in
later Tertiary and even Quaternary times. Further, during the
Pleistocene Period, there was very considerable movement of
sea level relative to the land owing to the vast quantities of

water locked up in the ice-sheets of the glacial periods and then discharged as the ice-sheets melted during the inter-glacial periods. This is in marked contrast to much of Upper Palaeozoic, Mesozoic and early Tertiary times, when the geological evidence gives little evidence of crustal instability and when warm temperate, if not tropical, climatic conditions seem to have been worldwide. On this view erosion at the present time is proceeding at an abnormally high rate, so that the present volume and sodium content of the rivers is well above average. A further fact to be considered is that the evaporite deposits have locked up part of the sodium carried into the oceans. Joly did not regard this loss as significant, but it must be remembered that nearly seventy years ago very few deep borings had been sunk, so that little was known of the nature and extent of beds that did not outcrop on the Earth's surface. As a result of the intensive exploration for oil during the past twenty years, we now know that large areas, such as the one to the south of the Great Lakes of North America, and another extending from northern Germany beneath the North Sea into north-east England, are underlain by beds of evaporite deposits up to several thousands of feet in thickness. The total sodium content of such evaporite deposits has recently been estimated as 400 million, million tons.

Joly calculated that 160 million tons of sodium were carried to the sea every year and that as the total quantity contained in sea-water was at least 90 million times greater, that the oceans had been in existence for about 90 million years. Sollas of Oxford, was at first inclined to reduce this figure to 30 million years, but in 1909 he increased it to 80 million years.

In 1963, D. A. Livingstone published a careful consideration of the sodium cycle. The latest data, based on a much more complete coverage of the world's rivers, raised the figure for the amount of sodium transported to the oceans to 205 million tons per annum with probable accuracy of 10–20%. In discussing how typical is the present of the past, Livingstone accepted Kuenen's estimate that on an average not more than one-fifth of the present land areas can have been covered by shallow seas during the quieter periods of the geological past. He found it very difficult to accept the suggestion that the carbon dioxide content of the atmosphere, and therefore the rate of chemical weathering of the rocks, is greater now than in the past. If the present is a time of abnormally high run off, then the age of the oceans determined by this geochemical method will be too young. Livingstone finally concluded that the present rate of weathering did not differ from the mean rate of the geological past by an order of magnitude.

Accepting the figure of 600 millions years based on isotopic data, as the duration of Phanerozoic times, Livingstone attempted to draw up a balance sheet between the amount of sodium carried to the seas by the rivers at present rates operating for 600 million years, and the sodium content of sea-water and the deposits on the ocean floor. On the one side he had the figure of $64 \cdot 2 \times 10^{15}$ tons of sodium transferred to the seas, on the other side he could account for only $24 \cdot 6 \times 10^{15}$ tons.

He argued that this considerable discrepancy must be due in part to the recycling of sodium that had been returned to the lands in rain after evaporation from the ocean surfaces, in part to the sodium present in sedimentary rocks, not only as obvious evaporite deposits, but also that which had been dissolved in pore water and absorbed to the surfaces of the particles of the finer-grained rocks, and in part to the sodium content of metamorphosed sediments.

There are so many uncertainties that Livingstone's final figures for the age of the oceans lie between 1313 and 2554 million years.

THE PALAEONTOLOGICAL 'SCALE'

In tracing the building up of the Stratigraphical Table, (p. 10) the importance of fossils was stressed. The fossiliferous units of the Table each yield their own distinctive assemblage of fossils: the faunas and floras yielded by each successive fossiliferous unit are different to a greater or less extent. As one works up through a fossiliferous succession, one notes that certain forms disappear, sometimes gradually, sometimes suddenly, whilst others take their place. Assemblages were not wiped out by catastrophies as was thought by certain of the earlier workers (p. 16), but they gradually evolved and merged into the slightly different assemblages found at higher stratigraphical horizons.

Much of the evidence for the reality of evolution in Charles Darwin's epoch making *Origin of Species* (1859) was taken from the palaeontological record. A century's further work has given us much more complete knowledge of the changes in forms of life during the geological past. At some horizons, such as the Chalk, we can trace in great detail the gradual revolutionary change shown by many groups of fossils, for example sea urchins, such as the heart-shaped micrasters. The Lower Tertiary beds of the western United States have yielded a superb sequence of specimens showing the earlier stages in the development of the modern horse, *Equus*, from its Eocene four-toed ancestor *Eohippus*, a swamp-dwelling form about the size of a modern fox terrier. But at other levels, as in the non-marine rocks of the Triassic System of many parts of the world, there are great thicknesses of beds, barren or almost completely barren of fossils. But despite such gaps in the record, which may well be filled by future discoveries, we now have a reasonably complete picture of the changing life of the past.

The younger Pre-Cambrian rocks have yielded the remains of a number of soft-bodied invertebrates, many of doubtful affinities, whilst in all the shield areas there are thick beds of limestone formed by lime-secreting marine algae. Invertebrates first appear in any abundance in the beds which are regarded as of early Cambrian age. The fauna of Lower Palaeozoic rocks is made up of corals, both solitary and colonial forms, primitive echinoderms, bivalved sessile brachiopods in great variety, together with the segmented trilobites and the graptolites, which are hemichordates. Fragmentary remains of primitive vertebrates and the first vestiges of land plants have been found at a few localities. In the Upper Palaeozoic strata one can trace the rise of the vertebrates (Fig. 14). Late Silurian and early Devonian times was the acme of the jawless agnathids, now represented only by the parasitic lampreys and hagfish; the bizarre and heavily armoured placoderms did not survive the

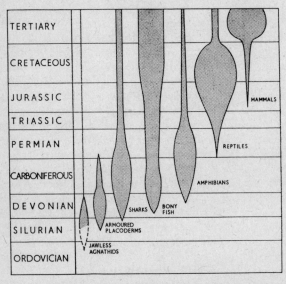

Fig. 14. *The history of the main groups of vertebrates. The variations in the width of the shaded areas indicate the periods of acme and of decline for each group.*

Palaeozoic. Of the other major groups of fish, the sharks and rays, with their jaws carrying great numbers of biting or crushing teeth and their skeletons made of cartilage, have not changed greatly since Palaeozoic times. This is in marked contrast to the bony fish.

Towards the close of the Devonian Period, amphibians developed from one group of the bony fish. They reached their acme in late Carboniferous and Permian times, a large variety of carnivorous and herbivorous types being known, but most of these died out during the Triassic Period. During the Carboniferous Period the reptiles evolved from certain amphibians and soon became the dominant vertebrate group. The Jurassic and the Cretaceous Periods are the 'Age of Reptiles' with the flying pterodactyls, the great swimming plesiosaurs and ichthyosaurs in the seas, and on land the dinosaurs, some nearly 100 feet in length and weighing many tons. But none of these 'ruling reptiles' survived into the Tertiary. The modern reptiles, crocodiles, snakes, etc., are all groups with a long geological history, during which they have changed but slowly. From the reptiles in late Mesozoic times developed the birds and, more important, the mammals. During the 'Age of Reptiles', mammals were small and insignificant, but early in Tertiary times, there was a great burst of mammalian evolution and development, culminating at the beginning of the Pleistocene Period with the appearance of man.

The second important feature of the Devonian rocks is the appearance of a land flora. At first only primitive marsh dwellers are known, but soon a varied land flora developed and by Carboniferous times flourished in sufficient luxuriance to form the coal seams of many parts of the world. The Upper Palaeozoic floras are made up partly of spore-bearing plants, partly of plants with unenclosed or naked seeds. Many of the dominant Upper Palaeozoic groups are now either extinct or very unimportant, others such as the conifers are mainly restricted to the areas of poorer soil. During the Mesozoic the spore-bearing plants decreased in importance and this was the acme of the gymnosperms with naked seeds. In late Mesozoic times the angiosperms (the flowering plants), with protected seeds, developed. They quickly supplanted the gymnosperms. Our modern flora in very large part dates back to the beginning of the Tertiary Era.

To return to the invertebrates (Fig. 15). At the beginning of the Devonian Period, the true graptolites died out, though the dendroid graptolites lingered on until well into the

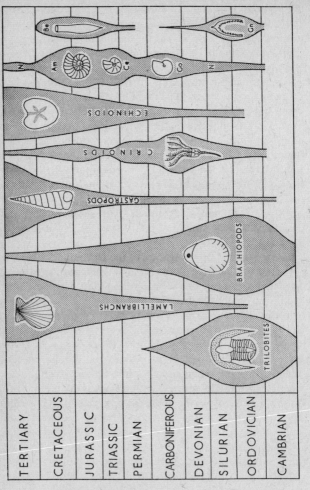

Fig. 15. *The history of the main groups of invertebrates. The variations in the width of the shaded areas indicate the periods of acme and of decline for each group.*
Am – *Ammonites*   Be – *Belemnites*   Ce – *Ceratites*
Gr – *Graptolithina*   Go – *Goniatites*   N – *Nautiloids*

Carboniferous. From highest Cambrian to Carboniferous times, the dendroid graptolites changed but little, in contrast to the rapid evolutionary changes undergone by the true graptolites, which make them of such value in the zonal subdivision of the Ordovician and Silurian rocks. In many other fossil groups one can see similar contrasts between some lines which were unprogressive and often very long-lived, others which show extremely varied evolutionary development over a relatively short time-span. The trilobites and the brachiopods, the two groups used for zoning the coarser-grained rocks of the Lower Palaeozoic, were both past their acme by Devonian times. The Devonian and the Carboniferous beds, particularly the latter, show a much greater variety of the bottom-living crinoids (sea lilies) whose remains often build up thick beds of limestone. Reef limestones built of corals and related forms are also well developed. From the infrequent nautiloids of the Lower Palaeozoic, with straight chambered shells sometimes of very large size, developed the coiled, usually smooth, goniatites which are often of zonal value in the marine clays of the Upper Palaeozoic.

On most continents the Triassic Period was mainly one of non-marine sedimentation, though where marine beds of this age are developed, as in parts of the Alps, the Triassic faunas are definitely Mesozoic (Middle Life) rather than Palaeozoic (Old Life) in character. Brachiopods were less important and varied, but the bivalved lamellibranchs and univalved gastropods were becoming more important. The goniatites were replaced by the ceratitic ammonoids with more heavily ornamented shells and greater complexity of the walls separating the chambers of their shells. From one line of the long-lived and slowly changing nautiloids had developed the belemnites with their strong pointed internal calcitic skeletons.

The corals of the Palaeozoic rocks had been replaced by other types of corals; most of the crinoids found in the Palaeozoic rocks had died out, but the echinoids (the sea urchins) were more abundant and varied. It is the richly fossiliferous rocks of the Jurassic and Cretaceous Systems which yield the typical Mesozoic faunas. These are characterized by the wealth and variety of the ammonites, which are of great zonal value, the acme of the belemnites, a limited range of brachiopods, but with both the lamellibranchs and the gastropods of considerably increased importance. Another change was in the echinoids. In the Palaeozoic rocks sea urchins are rare and all are regular forms with radially symmetrical tests. Regular echinoids persist throughout the Mesozoic rocks, but in addition there are a great variety of irregular echinoids with bilaterally symmetrical tests.

As a result of their greater variation in morphology, the Mesozoic sea urchins were able to live in a much greater range of sea bottom conditions than could their Palaeozoic ancestors.

Both the ammonites and the belemnites died out before the beginning of Tertiary times. Their close relative, the nautiloids, persist up to Recent times, as do the crinoids, but both these groups form extremely insignificant parts of the faunas found in the modern seas. The same remark applies to the brachiopods. The bottom living fauna of the Tertiary and Recent seas was dominated by the abundance and variety of lamellibranchs and gastropods and to a lesser extent of sea urchins, both regular and irregular, and corals.

We have traced in outline the changes shown by vertebrates, invertebrates, and plants for two reasons. Firstly, to show that palaeontologists base their extremely detailed subdivision of the units of the Stratigraphical Column on the recognition of fossils of zonal importance, that is forms with a very restricted time range and a wide geographical distribution. Secondly, that in contrast to such rapidly evolving groups as the ammonites and the true graptolites, there were other, often closely related, groups of fossils which changed very much more slowly.

Whilst the Stratigraphical Column can be subdivided on a palaeontogical basis into stages, and these into zones, this can only give relative order of age. Evolutionary change has clearly proceeded at different rates, both between the major groups of organisms and also between the different members of the same group. The two centuries or less of modern scientific investigation are too short a time-span to have witnessed, as far as we know at present, the evolutionary development of one species from another, except in the possible case of certain micro-organisms. The work of the geneticist and the animal and plant breeder has developed in many forms a wealth of hybrids and varieties, but they have all been derived from the interbreeding of members of the same species. It therefore follows that attempts to estimate the rate of evolutionary change in years and hence to provide an absolute palaeontologically controlled time-scale are entirely subjective.

For example, in 1867 Lyell published his estimate of the changes in the shelly fossils during the whole of the Tertiary Era as compared with the changes that had occurred since the beginning of the Pleistocene ice age. He concluded that the evolutionary changes during the Pleistocene were only one-twentieth of those that had taken place since the beginning of the Miocene Period. Assuming a duration of one million years for the Pleistocene, Lyell estimated the beginning of the Miocene Period at 20 million years ago. Extending the same method further back, Lyell estimated the beginning of the Tertiary Era at 80 million years ago and the beginning of the Palaeozoic at 240 million years ago. The first two figures argue reasonably well with the 26 and 65 million years respectively of the isotopic time-scale, but the last figure is less than half of the 570 million years now quoted.

Again, the American vertebrate palaeontologist, W. D. Matthew, working on the well-known Horse series, estimated that if the amount of anatomical change between the Recent *Equus caballus* and the early Pleistocene *Equus scotti* be taken as unity, then there were 85 of such units between *Equus caballus* and *Eohippus*, the first member of the series, found in beds of early Eocene age. Matthew assumed that the changes in the morphology of the Horse series had taken place at a steady rate, but this is by no means certain. Also at that time (1914) there were no means of estimating with any approximation to accuracy the duration of the Pleistocene Period.

LORD KELVIN AND THE RATE OF COOLING OF THE EARTH

By the middle of the last century, a younger generation of geologists had arisen to replace those who had first opposed and then in most cases reluctantly accepted the Doctrine of Uniformitarianism and its necessary corollary, the immensely long duration of geological time. The older catastrophic views were replaced by the ideas of gradual change taking place during virtually unlimited amounts of time.

But these views were attacked by certain physicists, notably by W. Thomson, later Lord Kelvin (1824–1907), who held in his later years a position of great prestige in British scientific circles. Kelvin thought that the geologists who placed no limits on the duration of time required for their Earth-history had ignored the principles of thermodynamics. He considered the Earth to be a globe which had cooled from an originally molten state; if the temperature of the initial molten state and the rate of subsequent cooling were known, then it would be possible to determine a limiting age for the creation of the world. He calculated the present rate of cooling of the Earth's surface from the available records of temperature gradients measured in mines, for the deeper the mine the higher the working temperature in it. In 1862, he published the results of his calculations in the *Philosophical Magazine*, a journal widely read by scientists, and also in *Popular Lectures and Addresses*. There were many uncertainties in his data, including the temperature of the original molten rock, which was thought to have ranged between 10 000° and 7 000° F. He concluded that the consolidation of the Earth must have taken place between 20 million and 400 million years ago. In later papers, he revised his figures, until in his last address on the subject in 1897, published in the *Philosophical Magazine* for 1899, he had reduced his limits to between 20 and 40 million years.

Despite Kelvin's great authority as a physicist, some geologists vigorously refuted his drastic shortening of the time at their disposal. For example, in 1897 J. S. Goodchild published in the *Proceedings of the Royal Physical Society of Edinburgh* a paper in which he drew up a time-scale based on an estimated rate of sedimentation. Goodchild regarded the base of the Cambrian as 704 million years old and wrote 'that the records of the rocks fully justify us in claiming for the Earth an antiquity so vast as to be far beyond the power of the human intellect to grasp'. Other geologists attempted to modify their thinking so that their requirements of time were not too dissimilar from Kelvin's figures. As will be shown later, the vast proportion of geological time represented by the Pre-Cambrian rocks had not then been appreciated, so when Sir Archibald Geikie in his Presidential Address to the Geology Section of the British Association in 1899 reluctantly agreed to accept 100 million years as the duration of geological time, only a small fraction of this was allotted to the Pre-Cambrian. This rather compromising attitude seemed to be supported by Joly's estimate in 1899 of 90 million years for the existence of the oceans, based on their sodium content (p. 46), whilst many of the estimates of the rate of sedimentation, such as Sollas's figures of 100 years for one foot (p. 38), seem to have been biased by the need to obtain a time figure which was not too discrepant from Kelvin's.

Kelvin in the true spirit of scientific caution had qualified his estimates with the proviso that they were correct 'only if sources of heat now unknown to us are not prepared in the great storehouse of creation'. The discovery of radioactivity not only demonstrated this to be the case, but, as will be shown in the next chapter, released geologists from the strait-jacket forced on them by Kelvin and provided for the first time the means for the accurate estimation of geological time.

5. Isotopic Dating

In 1895 Becquerel gave the name *radioactivity* to a phenomenon which he had noted with certain compounds of uranium. Even when wrapped in opaque material, they blackened photographic plates. Three years later Madame Curie and her husband Pierre Curie succeeded in isolating two intensely radioactive elements from the uranium-bearing mineral pitchblendes. One was named Polonium after Poland, Madame Curie's native country, the other Radium. Other spontaneously radioactive elements were soon discovered and now more than forty are known.

The radiations emitted were studied by Rutherford. In 1899 he concluded that they were of two kinds, which he termed α- and β-rays: a year later Pierre Curie recognized a third kind of radiation, γ-rays. Working in the Cavendish Laboratory at Cambridge, Rutherford showed, from their behaviour in magnetic fields, that the α-rays are positively and the β-rays negatively charged. Both kinds of rays consisted of particles. The α-rays of positively charged helium atoms moved at velocities ranging between 9000 and 12500 miles per second; the β-rays of negatively charged electrons and of neutrons not carrying a charge, travelled at velocities varying from the speed of light (184000 miles per second) to about a third of that rate. Later, in 1914, Rutherford and Andrade demonstrated that the γ-rays also moved with the speed of light and that they are very high frequency x-rays.

Rutherford also carried out fundamental work on the structure of atoms. In its simplest form, the hydrogen atom, it consists of a central nucleus made of one positively charged particle, the proton, round which orbits one negatively charged particle, the electron. The number of protons in the nucleus of an element determines its atomic number, e.g. 1 for hydrogen, 82 for lead, etc. But whilst the great majority of atoms of hydrogen are made up of one proton and one electron, about one in 5000 of the hydrogen atoms are heavier than normal. Investigation has shown that these 'heavy' hydrogen atoms contain, in addition to the protons and electrons, neutrons, particles of the same weight as the proton but not carrying an electrical charge. The total number of protons and neutrons in the nucleus of an atom determines its mass number. The 'heavy' hydrogen atoms with 1 neutron and mass number 2 are usually named deuterium and those with 2 neutrons and mass number 3 are named tritium. They have the same atomic number, 1, as hydrogen. Atoms with the same atomic number but different mass numbers having virtually identical chemical properties are described as isotopic. Atoms of one element with different mass numbers are termed isotopes of that element. Deuterium and tritium are therefore isotopes of hydrogen and are written H^2 and H^3, H being the chemical symbol for hydrogen. Two or more isotopes of the majority of the other elements are known, for example there are four isotopes of lead, Pb^{204}, Pb^{206}, Pb^{207}, and Pb^{208}.

Radioactive disintegration or decay results from the emission of either α or β particles from the complex atoms of radioactive elements. As mentioned above, these particles are moving at incredibly high velocities, so when they collide heat must be liberated. The heat produced by the decay of the radioactive minerals present in the rocks of the Earth's crust was unknown to Lord Kelvin (p. 57) and therefore falsified his calculations.

When the nucleus of a radium atom loses an α particle, which is a helium atom made up of 2 protons and 2 neutrons, it is changed into radon, a gas, and its atomic weight drops by 4, from 226 to 222. Radon is radioactive, so it will in its turn emit a helium nucleus and be transformed into another radioactive element, of atomic weight 218. The source atom is referred to as the mother and the derivative is the daughter atom. Thus radium breaks down by the emission of α particles to give daughter atoms of radon, and so on; this process will continue, until eventually a stable end product, in this case Pb^{206}, is produced. But radium itself is about halfway along a series or chain starting with uranium^{238} and ending with Pb^{206} (Fig. 16). During this series of transformations 8 helium nuclei (α particles) and a number of β particles have been emitted. Two other radiogenic isotopes of lead, Pb^{207}, and Pb^{208}, are the end members of two similar chains of radioactive transformations, which can be written:

Uranium$^{235} \rightarrow$ Lead^{207} + 7 α particles
Thorium$^{232} \rightarrow$ Lead^{208} + 6 α particles

In 1902 Rutherford and Soddy showed that the rate of decay of all radioactive elements obeyed a simple law. A single radioactive mother atom may break down instantaneously or be stable for any length of time. But when a large number of mother atoms are examined an average 'life expectancy' can be calculated. This average is found to be a constant however large the number of mother atoms may be. In a given unit of time (whether a year, a second or a century), the number of atoms, n, which decay, is directly proportional to the number of atoms, N, of the mother element. The fraction $\frac{n}{N}$ is termed the decay constant for the particular reaction and is denoted by λ. The decay constant for radium is 0·0004273, so that in an average year 4273 atoms will have disintegrated out of every 10 million atoms of radium present. No matter how large the sample, the length of time required for half the mother atoms to decay must be a constant. This is termed the half-life of the isotope. Fig. 16 shows the great variation in the length of half-lives in the various transformations in the $U^{238} \rightarrow Pb^{206}$ series, but in every case the decay constant and the half-life are related by the expression:

$$\text{decay constant} = \frac{0·693}{\text{half-life}}$$

The half-lives of the three radiogenic lead series are

$U^{238} \rightarrow Pb^{206}$	4510 million years
$U^{235} \rightarrow Pb^{207}$	713 million years
$Th^{232} \rightarrow Pb^{208}$	13 900 million years

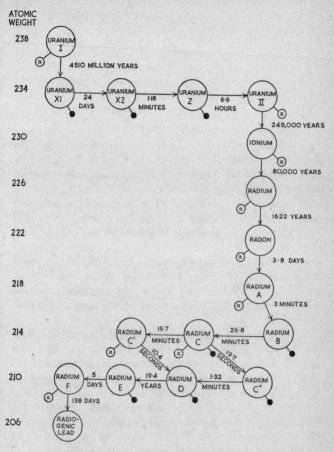

ATOMIC WEIGHT

238 — URANIUM I — α — 4510 MILLION YEARS

234 — URANIUM XI — 24 DAYS — URANIUM X2 — 1·18 MINUTES — URANIUM Z — 6·6 HOURS — URANIUM II — α — 248,000 YEARS

230 — IONIUM — α — 80,000 YEARS

226 — RADIUM — α — 1622 YEARS

222 — RADON — α — 3·8 DAYS

218 — RADIUM A — α — 3 MINUTES

214 — RADIUM C′ — 19·7 MINUTES — RADIUM C — 26·8 MINUTES — RADIUM B — α — 10·4 SECONDS — 19·7 SECONDS

210 — RADIUM F — 5 DAYS — RADIUM E — 19·4 YEARS — RADIUM D — 1·32 MINUTES — RADIUM C″ — α — 138 DAYS

206 — RADIO-GENIC LEAD

Fig. 16. *The U²³⁸/Pb²⁰⁶ Series. Open circles indicate the emission of an* **α** *particle, black circles of a β particle. The duration of the half-life of each phase is shown.*

The isotopic age of a rock or mineral containing a radio-active isotope which decays to a stable radiogenic end product can be calculated from the equation

$$T = \frac{1}{\lambda} \log_e (1 + \frac{D}{P})$$

where P = present day concentration of parent atoms

D = concentration of radiogenic daughter atoms

λ = decay constant of radioactive parent.

But we have gone rather ahead of the story of the discovery that radioactive elements can be used to measure time. This was first suggested by Rutherford in 1906. In the same year Strutt had detected the presence of radon in a wide variety of common rocks from different parts of the world, whilst in the following year Boltwood of Yale showed not only that lead is invariably present in all uranium minerals, but that the older the geological age of a mineral, the greater is its lead content. He also calculated the age of certain minerals and obtained figures ranging from 410 to 2200 million years. Shortly afterwards Strutt working on the accumulation of the other product of radioactive disintegration, the gas helium, obtained figures that were broadly comparable with Boltwood's.

At first many geologists, especially the older ones were very reluctant to accept such stupendous figures. One who did was Joseph Barrell of Yale, who wrote a number of extremely farsighted papers on many branches of geology before his untimely death in 1919, at the age of fifty. Another was the late Professor Arthur Holmes (1890–1965), who determined in Strutt's laboratory at Imperial College, London, the lead/uranium ratio of rocks which he had collected in Madagascar. Holmes's first paper on the measurement of geological time, published in 1911, was followed by a steady stream of papers and books, including several time-scales which he amended as more complete data became available. In 1964 the Geological Society of London published a Symposium on *The Phanerozoic Time Scale* and this was most appropriately dedicated to Professor Holmes.

For the earlier age-determinations only the lead and helium methods were known. These could only be applied to a limited range of minerals, mostly of considerable rarity. The results obtained also were not always consistent and naturally this encouraged scepticism. But when other methods of much wider applicability were developed, and these gave results which were in many cases consistent with those previously obtained, isotopic or radiometric dating became to be accepted generally as a most important tool. In addition, it made important contributions of its own, for it was able to provide definite answers to a number of unsolved geological problems, some of long standing.

REQUIREMENTS FOR ISOTOPIC DATING

The first requirement is that the radioactive elements or minerals which are to be measured must be present in sufficient quantity to allow measurements to be made. The quantities in which they normally occur are exceedingly small, usually only in terms of parts per million, so that the analytical techniques used have to be extremely refined. A great step forward was the development of the mass spectrometer, which greatly facilitated the determination of the minute amounts of the different isotopes present. The concentration of the elements and isotopes can now be measured down to a few parts per million with an accuracy of one to two per cent. The decay constants for the isotopes must be accurately known and also the initial mother and daughter ratio.

Secondly, as the method depends primarily on the proportion of daughter to mother element, it is necessary that the system under investigation should always have been a closed one. There are fourteen unstable nuclides with very varied half-lives in the chain U^{238} to Pb^{206}. As we have seen, one nuclide, radon, is a gas and therefore liable to escape, though as in this series its half-life is only 3·8 days, the loss may not be appreciable. But clearly, the danger of error is much greater if the age determination is based on the helium content of the specimen under analysis. Helium is the gas derived from the α particles. The emission of helium must have been spread over a very long period of time, so the possibility that part, perhaps an appreciable part, of the helium will have been lost by leakage is very considerable. Moreover many minerals are liable to be attacked and altered, either by hydrothermal changes soon after their formation, or by selective weathering when they have been uplifted into the zone of groundwater. If there has been differential leaching or selective removal in other ways, then, clearly, inexact dates will be obtained. Often prolonged search and the expenditure of much physical energy in breaking up rock is necessary to obtain samples sufficiently fresh to be worth analysing.

Finally, if the isotopic age determined is to be tied into the Stratigraphic Table, it is necessary that the relative geological age of the sample should be known within narrow limits. A considerable number of precisely dated points are clearly necessary before a stratigraphical succession can be converted into an exact time-scale.

Before discussing the results obtained by isotopic age determinations, it will be necessary to outline the several methods used and in each case to indicate possible limitations.

THE URANIUM-THORIUM-LEAD METHOD

Most uranium minerals contain some thorium and most thorium minerals contain some uranium. It is therefore normal practice to analyse for the three elements, thorium, uranium and lead; the earlier workers determined total uranium, thorium and lead by gravimetric methods and then the atomic weight of the separated lead by direct weighing. These extremely laborious chemical separation methods have now been supplanted by mass-spectrometric measurements; the isotopic composition of the lead is determined in terms of the isotopes Pb^{204}, Pb^{206}, Pb^{207} and Pb^{208}. We have already seen that the last three isotopes are radiogenic, but there is no evidence that Pb^{204} has a radioactive source. Its presence shows that the radiogenic lead is 'contaminated', for some common lead must have been enclosed in the radioactive lead when the mineral was formed. (In isotope studies, three types of lead are recognized. First *primaeval* lead that is assumed to have existed at an early stage of the Earth's history, then *common* lead which differs from primaeval lead in containing *radiogenic* lead that is produced by the radioactive decay of uranium and thorium.) A correction must be applied for this, otherwise the age determined will be in error, and this is done by determining the isotopic composition of common lead formed in the same area and at the same time as the radioactive material. The percentage amounts of the radioactive isotopes of lead found in the common lead must therefore by subtracted from the proportions found in the analysed radioactive lead.

We therefore have a number of ratios, $\dfrac{Pb^{206}}{U^{238}}, \dfrac{Pb^{207}}{U^{235}}, \dfrac{Pb^{208}}{Th^{232}},$ for determining age, provided that the decay constant (p. 60) is known correctly in each case. In addition, because of the very different rates at which U^{238} and U^{235} decay, the $\dfrac{Pb^{207}}{Pb^{206}}$ ratio is of value in age-determination in the older rocks. Fig. 17 shows the value of these ratios for the period of the last 600 million years and also possible errors in the decay constants. Clearly for a relatively short period like this, any analytical errors affecting the $\dfrac{Pb^{207}}{Pb^{206}}$ ratio will have serious results. In an uranium-bearing mineral that crystallized during this time-span, Pb^{206} can be determined with much greater precision than Pb^{207}, for it will form about 95% of the uranium-lead. Therefore the most reliable ratio is likely to be that of $\dfrac{Pb^{206}}{U^{238}}$.

If, as in the example below (the pitchblende from Joachimsthal, Czechoslovakia, used by Mme Curie for the first isolation of radium), the three ratios are consistent, then one can accept the isotopic age-determination with confidence.

Pb^{206}/U^{238}	244 million years
Pb^{207}/U^{235}	249 million years
Pb^{208}/Th^{232}	242 million years

But the results obtained are not always so consistent; for example, the analysis of a Canadian sample of pitchblende gave

Pb^{206}/U^{238}	337 million years
Pb^{207}/U^{235}	389 million years
Pb^{208}/Th^{232}	705 million years

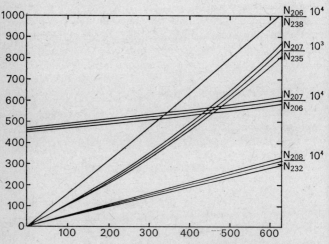

Fig. 17. *Numerical values of uranium-thorium-lead ratios plotted against age.* N *is the number of atoms of the various isotopes. The central line shows the mean value of the ratio; The outer lines show the limits of uncertainty of ratio owing to the uncertainty with which decay constants are known. Note that for* N^{206}/N^{238} *the uncertainty is within the width of the line.* (Reproduced with permission from The Phanerozoic Time-Scale, Geological Society Symposium, 1964, p. 75.)

showing that this mineral must have been affected by leaching or other change since its time of formation. With such discordant results it is impossible to give a reliable age-determination. These figures are rather unusual, for they clearly show thorium loss. It is often the lead/thorium age that is lower than those given by the lead/uranium ratios.

Pitchblende, uraninite, monazite, zircon and thorianite are the commonest uranium- and thorium-bearing minerals, which can be investigated by this method. The average concentration of thorium and uranium in the rocks of the Earth's crust is extremely low. For uranium it is about 4 parts per million in igneous and 2 parts per million in the sedimentary rocks. Very locally, however, one may find mineral veins or dyke-like masses of igneous rock carrying appreciable concentration of radioactive minerals, such as pitchblende, uraninite or radiogenic varieties of the common lead ore, galena (lead sulphide). The isotopic age of such bodies can be determined, but too often it cannot be related at all precisely to the geological scale. For instance, two suites of dykes are shown in Fig. 4, one cut off by and therefore older than the plane of unconformity, the other cutting the rocks above the unconformity. The older dykes are clearly post-the Folded Series–pre- the Bedded Succession, whilst all that can be said about the younger dykes is that they are post- the Bedded Succession. If the geological ages of the Folded Series and of the Bedded Succession are very different, then there is clearly a long time gap during some unknown part of which the older dykes were emplaced. The extent to which the other dykes are younger than the Bedded Succession is equally unknown.

Zircon and monazite both occur as accessory minerals, that is, minerals normally present in very small amounts, in igneous rocks, particularly in granites. Locally they may be present in more than the normal concentrations and can be abstracted by suitable techniques. But again it may be impossible to give more than an extremely approximate geological date to the emplacement of a granite batholith. The batholith shown in Fig. 4 is post- the Folded Series, for it cuts and alters them. Both the younger dykes and the volcano are shown as connected to the granite, but they may be much later than the granite batholith and have been produced by a later recrudescence of igneous activity in the area. If the same granite mass is exposed elsewhere at the Earth's surface, so that it can be sampled, then isotopic determinations from the younger dykes and the granite will show whether these were emplaced at approximately the same or different periods of time.

Zircon and monazite are both tough resistant minerals, so that if their host rock is exposed to erosion, they may well survive to be incorporated in sediments as detrital grains, often well worn and rounded owing to the abrasion they have suffered. In certain places, as at Prado, Brazil, at Travancore, India and in Ceylon, the beach-sands are unusually rich concentrates formed from the erosion of monazite-bearing granites. The isotopic ages obtained from the mineral grains in such detrital deposits will be those of the host rock in which the minerals originally crystallized and not those of the deposits in which they now occur.

Certain sedimentary rocks have an unusually high uranium content. The best known and richest are the nodules and seams of Kolm in the Alum Shales of Sweden. The average content of U_3O_8 in the Kolm is nearly half a per cent, but as the Kolm is scattered irregularly through the Alum Shales, the uranium content of the Alum Shales as a whole is far below this figure. The palaeontological age of Alum Shales (Upper Cambrian) is precisely known, but unfortunately the age-determinations made show considerable discrepancies as in the example below:

	U^{238}/Pb^{206}	U^{235}/Pb^{207}	Pb^{207}/Pb^{206}
Edge of nodule	365	435	920
Centre of nodule	370	425	770
Edge of nodule	255	435	945

These discordant ages are probably due to removal of radiogenic lead, which was enriched in Pb^{206} as a probable consequence of diffusion of the gas, radon. There has also been controversy as to whether the uranium-rich hydrocarbons were formed at the time of deposition or as a result of

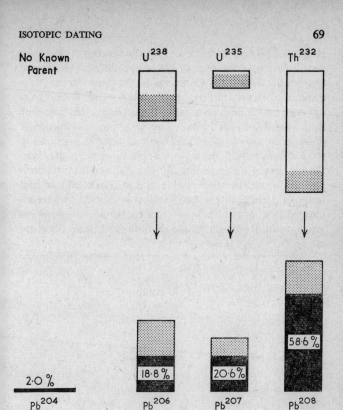

Fig. 18. *Relations between uranium, thorium and lead isotopes. Black – primaeval lead (proportion in the Canyon Diabolo meteorite). Grey – radiogenic lead produced by the decay of uranium and thorium in 4 500 000 000 years. White – unchanged uranium and thorium. (After R. S. Cannon Jr. and others.)*

diagenetic change some time, probably no great time, after the deposition of the Alum Shales.

THE COMMON LEAD METHOD

Lead occurs not only in the lead ores, e.g. galena, lead sulphide but also in very small amounts, a few parts per million, in certain of the minerals, e.g. the micas and the

potassium feldspars, that build up the igneous rocks. Very
detailed isotopic analysis of galenas of differing geological
age have shown slight variations in the proportions of the
three radiogenic isotopes Pb^{206}, Pb^{207}, and Pb^{208}. It is
assumed that at an early stage of the cooling of the Earth, all
lead was of the same isotopic composition and that these
differences from the primordial or primaeval lead are due to
the addition of radiogenic lead. The lead of meteorites containing negligible amounts of thorium and uranium is
regarded as representing primaeval lead. The total lead content of the rocks of the Earth's crust is thought to have increased by approximately 20% during the past 3000 million
years owing to the formation of radiogenic lead. Since the

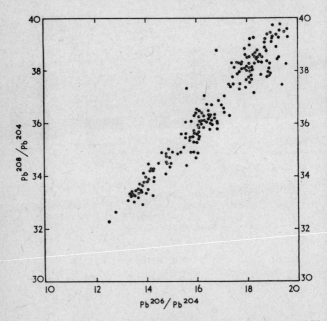

Fig. 19. *Plot of analyses showing ratios of radiogenic Pb^{206}
and Pb^{208} against nonradiogenic Pb^{204}. (Data from R. D.
Russell and R. M. Farquhar,* Geochim cosmochim. Acta,
1960.)

time the Earth was formed, 50% of the original U^{238} has decayed to form Pb^{206} and 90% of U^{235} to form Pb ^{207} (Fig. 18).

A plot of analyses showing the ratios of radiogenic Pb^{206} and Pb^{208} against the non-radiogenic Pb^{204} falls almost along a straight line (Fig. 19). Therefore, if the decay constants are known, the ratios of radiogenic to non-radiogenic lead can be used for dating purposes. The dates obtained, however, are not those of the crystallization of the lead-bearing minerals, but of the time when the lead separated from its source deep beneath the Earth's surface. For instance, a mineral vein in Perthshire contains lead, zinc and uranium minerals, which appear to be contemporaneous on geological evidence. The age of the pitchblende is 230 million years, whilst the lead determinations are of the order of 550 million years.

Age-determinations by the common lead method therefore need very careful interpretation and should be compared, if possible, with those obtained by other techniques. It is possible that mineralization may have extended over a considerable period of time, successive mineralizing fluids moving through the same network of veins. Indeed, considerable lead isotope variations have been found between the successive layers or growth zones of a single crystal of galena.

THE HELIUM METHOD

Strutt's original age-determinations (p. 62) were made by determining the thorium, uranium and helium content of certain minerals and calculating their ages in terms of the rate of production of helium from the parent thorium and uranium. But whilst all the ages obtained fell in the correct geological order, they were all considerably younger than those determined from rocks of corresponding geological age by the lead methods; clearly a part of the helium generated had leaked away and therefore the ages obtained were not the true but only minimum ages.

Despite very considerable improvements in analytical techniques, so that amounts as little as one-millionth of a cubic centimetre of helium can be measured, other discrepancies have been found. The possibility of helium loss is always present and therefore this method is not now used.

Since the end of World War II, the original lead and helium methods have been very largely replaced by the rubidium/strontium (Rb/Sr) and potassium/argon (K/Ar) methods, which have the advantage not only of somewhat simpler analytical procedures, but more important, of being applicable to a much greater range of minerals and rocks.

THE RUBIDIUM-STRONTIUM METHOD

There are two isotopes, Rb^{85} and Rb^{87}, of rubidium. Rb^{87} is radioactive and, by the emission of a β particle, changes to one of the isotopes of strontium, Sr^{87}. This method was first suggested in 1938 by Strassmann and Walling after they had isolated radiogenic strontium from a rubidium-bearing mica in a pegmatite from Manitoba. The pegmatite had an age of about 2000 million years by the lead method. The strontium isolated proved to be nearly pure Sr^{87} as compared with the seven per cent of Sr^{87} found in ordinary strontium, which contains three other isotopes.

In 1946 Ahrens employed the method extensively in determining the age of South African rocks, and since then the techniques involved have been improved by a number of workers. One difficulty has been the determination of the half-life of Rb^{87}. Strassmann and Walling calculated this as $6 \cdot 3 \times 10^{10}$ (63 000 million) years, but later work showed this to be too high and now most workers accept the figure of $4 \cdot 7 \times 10^{10}$ years. Age-determinations of a number of micas by both the K/Ar and the Rb/Sr methods have given extremely consistent results when the $4 \cdot 7 \times 10^{10}$ half-life is used. Calculations made using a longer half-life value give younger ages, but this can be corrected.

The igneous rocks are composed of a limited number of

rock-forming minerals. These comprise quartz, silicon
dioxide, and a number of complex silicates, some light,
others dark in colour. The light-coloured silicates are the
whitish or pink feldspars and certain micas, such as muscovite and lepidolite. The feldspars comprise potassium-rich
types, such as orthoclase and microcline and the soda-lime
plagioclases. The coloured minerals consist of the dark
micas, such as biotite and phlogopite, the amphiboles (e.g.

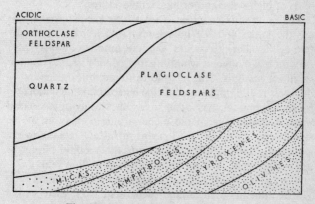

Fig. 20. *The variation in the relative abundance of the
chief rock-forming minerals in the igneous rocks. Coloured
minerals are stippled.*

hornblende), the pyroxenes (e.g. augite) and the olivines.
Igneous rocks vary considerably in their chemical composition, but a convenient broad distinction is into the light-
coloured acidic types, such as granite, and the dark heavier
basic rocks, such as basalt. One easily recognizable feature
of the acidic rocks is the presence of 'free', that is visible,
quartz. This is absent from the basic rocks with their much
lower silica content, for in them the silica is combined with
other elements to form the silicate minerals. The igneous
rocks are divided on a basis of textural features and detailed
mineralogical composition into a large number of rock-types,
grading in the coarse-grained types from acid granite
to basic gabbro and amongst the lavas from basic basalt to

acid rhyolite. The relative distribution of the different rock-forming minerals amongst igneous rocks is shown in Fig. 20.

The metamorphic rocks are composed of much greater variety of minerals than are the igneous rocks. Their constituents may include quartz and the same silicates as in the igneous rocks, with the exception of the olivines, whilst the pyroxenes are much less common. In addition they may contain a number of other silicates, which form under conditions of high stress and/or high temperatures.

The most suitable minerals for Rb/Sr age-determinations are the micas and the potassium-rich feldspars, especially microcline. The pyroxenes and the amphiboles may also contain rubidium and strontium, but usually in amounts too small for analysis. This method can therefore be applied to minerals found in the more acidic igneous and metamorphic rocks. Such rocks may occur in a wide variety of geological settings, ranging from the metamorphic complexes of the shield areas to discordant igneous intrusions. Owing to the extremely long half-life of rubidium, this method cannot be applied to rocks younger than about ten million years.

The earlier Rb/Sr age-determinations were made on individual minerals painstakingly separated from their parent rock, but recently there has been an increasing tendency to use as well 'whole-rock' analyses carried out on specimens 4–6 lbs in weight. Radioactive changes may have moved beyond the confines of a single crystal grain. It is often found that if whole-rock as well as individual mineral determinations are made on the same specimen, the whole-rock age > microcline age > muscovite age > biotite age. It is therefore desirable when giving an Rb/Sr age to state precisely what has been analysed.

ISOCHRONS

If a whole rock is analysed for Sr^{87} and Rb^{87}, the total Sr^{87} present is the sum of the original Sr^{87} (that present at the time of formation of the rock or mineral) and that developed subsequently from the decay of Rb^{87}.

$$\text{Sr}^{87} \text{ now} = \text{Sr}^{87} \text{ (initial)} + \text{Sr}^{87} \text{ from Rb}^{87}$$
$$= \text{Sr}^{87} \text{ (initial)} + \text{Rb}^{87} \, (e^{\lambda t} - 1)$$

The Sr^{86} content is also determined, for Sr^{86} is not radiogenic. Hence it can be used as a control. If the equation above is divided by Sr^{86}, for all but the oldest rocks.

$$\frac{\text{Sr}^{87}}{\text{Sr}^{86}} \text{ (total)} \approx \frac{\text{Sr}^{87}}{\text{Sr}^{86}} \text{ (initial)} + \frac{\text{Rb}^{87}}{\text{Sr}^{86}} \, \lambda$$

The results of a number of whole-rock analyses carried out on specimens from the rock mass are then plotted on a diagram with $\dfrac{\text{Sr}^{87}}{\text{Sr}^{86}}$ (total) graphed against $\dfrac{\text{Rb}^{87}}{\text{Sr}^{86}}$. The different analyses should fall on a straight line, an isochron (Fig. 21) whose slope is proportional to the age of intrusion, whilst its intercept with the axis of the graph gives the original $\dfrac{\text{Sr}^{87}}{\text{Sr}^{86}}$ ratio of the rock. But if subsequent to its emplacement, the intrusion has been subjected to metamorphism, its constituent minerals will have been differently affected. Analyses of micas and potash feldspars plotted on the same diagram should fall on an isochron of different slope (Fig. 21), and this will give the age of the metamorphic

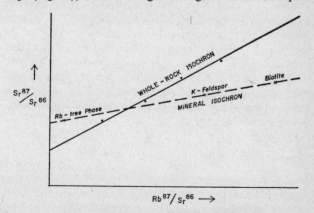

Fig. 21. *Whole-rock and mineral isochrons for a metamorphosed granitic rock. (After Moorbath, 1965.)*

event. For example, the isochron for whole-rock Rb/Sr
analyses from the Carn Chuinneag intrusion in Scotland (Fig.
29 and p. 105) yielded an age of 530 ± 10 million years with
an original Sr^{87}/Sr^{86} ratio of 0.710 ± 0.002. The feldspar-
biotite-muscovite isochron gave an age of 390 ± 5 million
years with an Sr^{87}/Sr^{86} ratio of 0.782. The 530 million year
date is interpreted as the age of the intrusion, the 390 as that
of a subsequent metamorphic 'event'.

But not all analyses have yielded such sharply defined
isochrons as in the case cited above. Complications will be
caused if the system has not remained a closed one, especially
if rubidium has been added to the rock during metamorphism.

In certain orogenic belts it has been found that U/Pb determinations on zircon have yielded considerably older ages
than those obtained by the Rb/Sr and K/Ar methods. For
example, zircon ages of 2700 million years as compared
with Rb/Sr and K/Ar ages of 1800 to 1900 million years have
been determined from the metamorphosed Pre-Cambrian
rocks of parts of Finland. The zircon ages relate to an
earlier cycle of high-grade metamorphism and mineral formation. The Rb/Sr and K/Ar to a later one which had
affected the same area. Evidence of the age of the first
metamorphism had been preserved only in the resistant
zircons. In other minerals it had been overprinted by
the second metamorphism and it was this age that they
recorded.

The whole-rock Rb/Sr method has recently been success-
fully applied to samples of shale and siltstone, the slope of
the isochron giving, in unmetamorphosed sediments, the age
of their deposition.

THE POTASSIUM-ARGON METHOD

The potassium-argon method of age-determination first
suggested in 1948 by L. T. Aldrich and A. O. Nier is of
wider applicability than the other methods. There are three
isotopes of potassium, K^{39}, K^{40} and K^{41}.

K^{40} is radioactive and undergoes a double transformation, radiogenic calcium (Ca^{40}) being produced by the emission of a β particle and radiogenic argon (Ar^{40}) by a more complex process involving both electron capture and the emission of γ-radiation. It is extremely difficult to distinguish radiogenic Ca^{40} from the much more abundant natural Ca^{40}, so this method relies on the measurement of the potassium by flame photometry or by neutron activation analysis and of the radiogenic Ar^{40} by isotope dilution analysis. The half-life of K^{40} is about 1300 million years, much less than that of rubidium[87], so that whilst the potassium method can be used like the rubidium method for Pre-Cambrian rocks with ages of several thousand million years, it has the advantage of a considerably lower minimum age, less than one million years in place of ten million years.

Potassium in determinable amounts occurs in a much greater range of rock-types than does rubidium. In igneous and metamorphic rocks, there are the potassium-bearing micas, biotite, muscovite, phlogopite and lepidolite, the potassium-bearing feldspars, orthoclase, microcline and sanidine, whilst the potassium-bearing pyroxenes and the amphiboles can be determined, though with greater difficulty for their potassium contents are so low. It will thus be seen (Fig. 20), that this method can be applied to a greater range of acid to basic rocks than can the rubidium method. Rather surprisingly it has been found that despite their very well developed cleavage, micas have greater argon retention than the less well cleaved feldspars. Therefore nowadays micas are analysed in preference to feldspars. Whole-rock analyses are also made.

A number of potassium-bearing minerals are found in sedimentary rocks. When the method was first developed, there were high hopes that accurate determinations of the isotopic age of fossiliferous sediments would enable direct correlation to be made between the isotopic and the geological scales, but unfortunately the more thorough trial of the method has been rather disappointing.

Glauconite is a complex potassium-bearing silicate, which

is being formed today on parts of the sea floor off the coasts
of California, where deposition is proceeding very slowly.
Geological evidence shows that in the past glauconite was
formed under similar conditions, conditions that are rather
specialized and unusual. In the British Stratigraphical
Column, glauconite-bearing sediments are to be found in the
Lower Cambrian rocks of the Midlands and after that not
until the 'greensands' of highest Jurassic and Cretaceous
ages and in marine rocks of the Eocene. In other countries,
glauconitic sediments occur at many intervening horizons
and hence fill the gaps in the British record. Unfortunately
glauconite is a mineral that is particularly prone to the
effects of weathering, whilst too deep burial beneath the
weight of later deposited sediments may also lead to loss of
argon. The majority of the ages determined from glauconites
are younger than those obtained from beds of comparable
age by other methods. This is most unfortunate, for most
glauconite-bearing beds are fossiliferous. It seems that the
best that can be said for glauconite dates is that they give
minimum, but not necessarily exact, ages to sediments.

Amongst the clay minerals (p. 32) are a number of fine-
grained micaceous complex silicates forming the Illite group.
Certain illites have yielded results which are in good agreement with their probable age, but more detailed investigations have shown that this may be fortuitous. Illite-bearing
shales have been thoroughly disintegrated and K/Ar age-
determinations were then made on the different size fractions. The coarser fractions gave ages considerably older
than the probable one, the finer fractions considerably
younger. The coarser fractions contained detrital muscovite,
whilst the illite that formed the bulk of the finer fractions
was thought to be a poor retainer of argon.

Certain evaporite deposits contain sylvite (potassium
chloride), but age-determinations made on them have been
anomalous. Evaporite deposits are especially prone to diagenetic modification.

Another unusual rock-type which may be found interbedded with normal sediments is bentonite. Bentonites are

extremely fine-grained volcanic ashes, the products of explosive eruptions some distance from the place of sedimentation. Bentonites may contain potassium-bearing minerals such as sanidine, biotite, together with feldspars of the plagioclase series with only a low potassium content. Age-determinations of bentonites, particularly on sanidine and biotite, are likely to be more accurate than those on glauconites.

THE RADIOCARBON OR CARBON[14] METHOD

The radiocarbon or C^{14} method of dating devised in 1947 by W. F. Libby of Chicago, has been of great assistance to the archaeologist and those geologists who are concerned with the extreme top of the Stratigraphical Table.

Neutrons produced by cosmic ray flux in the outer atmosphere collide with nitrogen atoms to produce the radioactive isotope of carbon, C^{14}. Like ordinary carbon this is oxidized to carbon dioxide and helps to swell the vast quantities of carbon dioxide held in the atmosphere and in the waters of the hydrosphere. Therefore the carbon dioxide assimilated by organisms, particularly by plants, will include an exceedingly minute amount of radiocarbon. But acquisition of radiocarbon will cease at the organism's death, after which the radiocarbon will follow its normal pattern of decay. It is generally accepted that the production rate of radiocarbon from cosmic flux has been substantially constant for many thousands of years. If this was not the case, the results obtained by this method would surely be anomalous. Recent work has shown that there have been small fluctuations during the past 6000 years, but these can be allowed for.

The half-life of radiocarbon is only 5730 years, so that material older than a few half-lives contains so little radiocarbon (Fig. 22) that its measurement becomes extremely difficult. Material older than 50000 years contains only a fraction of a per cent of the minute amount originally present. Determinations have been made for ages up to 70000 years, but using the most refined techniques and

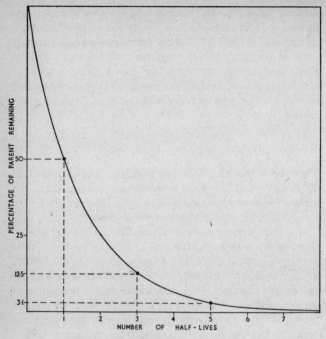

Fig. 22. *The half-life curve.*

equipment available there is then a considerable experimental error.

Radiocarbon dating can therefore only be carried out on material collected from deposits of Holocene and very latest Pleistocene age. One possibility that has to be rigorously guarded against is contamination by modern organic matter. The material to be analysed may have been penetrated by tree roots or soaked by water rich in manure or sewage, packing materials must be carefully selected and scrupulous cleanliness observed in the laboratory. If contamination has occurred the measured age will be too young and the older the specimen the larger will be the discrepancy between the true and the 'measured' age. Samples should be treated with sodium hydroxide solution as this will extract

modern humic matter together with some of the original
material. Discrepant ages obtained from the original
material and the humic extract are a warning of contamination and that the measured age is an underestimate. Dates
obtained from shells and bones are particularly liable to
inaccuracy. Percolating water will not only attack the
carbonates, but isotope replacement is likely to occur
between C^{12} and C^{14}, causing enrichment of the latter.

There are, however, so many radiocarbon dates telling a
consistent story that this method has made major contribution to our understanding of the events of the past 50 000 or
so years.

As Fig. 23 shows, there is an urgent need to develop a new
method to fill the large gap between range of ages determin-

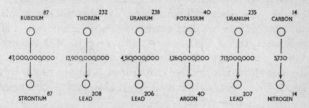

Fig. 23. *Half-lives of the main isotopic minerals used in
geochronology.*

able by the C^{14} and the potassium-argon methods. Beryl-
lium[10], another product of cosmic-ray bombardment,
emitting β particles with a half-life of 2 500 000 years, has
been suggested, but it is such a rare constituent of rocks and
minerals that, even if the analytical difficulties were to be
overcome, its use would probably be distinctly limited. A
number of other techniques, such as ionium/thorium,
protoactinium[231]/thorium[230], have been developed very
recently, but the results are not yet consistent enough to
allow the suggested dates to be accepted with confidence.
The dating of minerals, such as zircon, by the study of the
tracks produced by the spontaneous fission of U^{238} is
another new technique, whose potentialities and limitations
have yet to be evaluated.

6. The Isotopic (Radiometric) Time-Scale*

THE DEVELOPMENT OF THE PHANEROZOIC SCALE

Arthur Holmes was the architect of the earlier time-scales. In the first (1913) edition of his well-known book, *The Age of the Earth*, he published a graph (Fig. 24) showing that three U/Pb age-determinations from igneous rocks whose stratigraphic positions were approximately known, fell nearly in a straight line when plotted against the maximum thickness of sediments. In the 1937 edition, with the help of additional U/Pb and U/He determinations, Holmes published a detailed time-scale for the Phanerozoic. Ten years later this had to be revised fairly considerably. The mass spectrometer

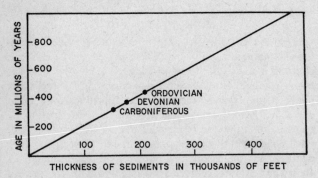

Fig. 24. *The first plot of isotopic age against thickness of strata. (After A. Holmes, 1913.)*

*Following the practice of the Geological Society, the term isotopic is adopted here in place of the older term radiometric.

had by then come into use; the isotopic composition of the lead could be determined with much greater precision than hitherto, so the accuracy of the ages on which the 1937 time-scale was based were now in question. In 1947, in a well-known paper in the *Transactions of Geological Society of Glasgow*, Holmes put forward his revised time-scale, showing two alternatives. He based his time-scale on five recently made determinations, but whilst their analytical accuracy was high, in three cases there was doubt as to the precise stratigraphical position. The 58 million year determination might be assigned either to the late Cretaceous or to the early Palaeocene, the 214 million year one to the early Permian or to the late Carboniferous and the 255 million figure either to the Lower Carboniferous or to the late Devonian. The two earlier ages were more precisely delimited. Holmes therefore drew two curves (Fig. 25), using for the one the earlier and for the other the later of the possible datings. The time-span of the periods was read off from the curves, using thickness of sediments as the index of their relative duration. Curve B was thought to be the more probable and most geologists used this scale for the next ten years. In 1959 Holmes revised his time-scale taking into account many new dates, not only by the older U/Pb and Th/Pb methods, but also those which had begun to appear based on the K/Ar and Rb/Sr methods. As shown in Table 1, the durations of the Palaeozoic systems were considerably expanded. The dates given were stated as ± 5 or more million years, with the beginning of the Cambrian period as 600 ± 20 million years ago. In 1961 J. L. Kulp put forward his time-scale, based on many new determinations. As shown in Table 1 there is, except in a few places, close agreement with Holmes's 1959 scale. The latest (1964) Phanerozoic time-scale appeared in the symposium dedicated to Professor Arthur Holmes and published by the Geological Society of London. This time-scale is the most comprehensive yet produced, for it gives a suggested age not only for the beginning of each period, but also for each stage into which the period has been subdivided. The volume contains critical discussion from the majority of the

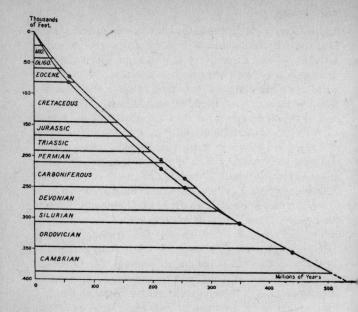

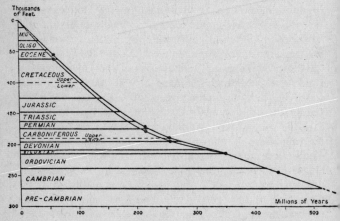

Fig. 25. *A Holmes* 1947, A *(Upper) and* B *(Lower) curves. (Reproduced, with permission, from* Trans. Geol. Soc. Glasgow, 1947, **21**, *p. 143.)*

contributors as to the accuracy of the data used for the erection of the time-scale.

Table 1. *Estimates of the duration of the Phanerozoic Periods (in millions of years)*

	Holmes (1937)	Holmes (1947) A	Holmes (1947) B	Holmes (1959)	Kulp (1961)	Symposium (1964)
Pleistocene	1	1	1	1	1	1·5–2
Pliocene	15	14	11	10	12	c5
Miocene	16	17	14	14	12	c19
Oligocene	16	15	12	15	11	11–12
Eocene and Palaeocene	20	21	20	30	27	27–28
Cretaceous	40	72	69	65	72	71
Jurassic	37	27	25	45	46	54–59
Triassic	48	29	30	45	c49	30–35
Permian	34	24	21	55	50	50
Carboniferous	48	55	52	80	65	65
Devonian	38	45	58	50	60	50
Silurian	28	32	37	40	c20	35–45
Ordovician	51	80	80	60	c75	60–70
Cambrian	78	80	80	100	c100	c70

Note: *All the dates in Holmes's 1959 Table from the Cretaceous downwards are mean figures.*

Before discussing the Phanerozoic time-scale in detail, it will be as well to consider the basis on which it has been constructed. The number of technically reliable dates is now considerable. In the 1964 symposium 337 such determinations are listed. The great majority (280) are by the K/Ar method, 29 by Rb/Sr and the remainder by the older (usually U/Pb) methods. As already mentioned (p. 72), there has been some disagreement in the past as to the precise value of the decay constants, so where necessary ages have been recalculated using standard decay constants. Most determinations are based on a number of analyses of material from the same locality. In some cases determinations have been made on more than one mineral, e.g. K/Ar on biotite and sanidine, in other cases whole rock has been used.

The mean age for each locality can therefore be stated with precision including the probable error, e.g. the date of intrusion of the Dartmoor granite is given as 295 ± 5 million years.

The uncertainty, all too often, is in the exact stratigraphical horizon of the material analysed. For instance, the Dartmoor granite cuts sediments of Lower Carboniferous

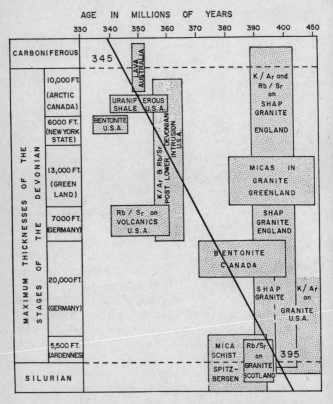

Fig. 26. *Age-determination data for the Devonian System. The size of the boxes shows the possible isotopic and stratigraphic errors for each determination. The line of 'best fit' is thickened. (After M. R. House in* The Phanerozoic Time-Scale, Geological Society Symposium, 1964.)

age. It was exposed to erosion at some time in the Permian, but the Permian sediments of Devonshire and Somerset are completely barren of fossils and therefore cannot be dated at all precisely. It can be argued on the balance of geological probabilities, that the Dartmoor granite was intruded towards the close of the Carboniferous Period, most probably in the Stephanian Stage, but the evidence is too slender to use this well-established isotopic age, as was done by Holmes in some of his earlier attempts, as definite evidence for fixing the Permian/Carboniferous boundary. The same difficulty applies in a greater or lesser extent to the great majority of dates determined on intrusive igneous rocks.

Fig. 26 shows the way to recognize and allow for the limitations of the best data at present available. Full allowance is made for the possible errors, both in the accuracy of the age-determinations and also in the stratigraphical position of the specimens analysed. For vertical scale the maximum-known thicknesses of the various subdivisions of the system are used, even though these may have been measured in widely separated areas. A line of 'best fit' is then drawn, taking into account the most probable stratigraphical age of the igneous rocks. In the example given, the extreme age limits of the emplacement of the Shap Granite are post-Upper Silurian, for it metamorphoses strata of this age, and pre-Lower Carboniferous, for pebbles of Shap granite are to be found in the near-by Lower Carboniferous conglomerates, although on more indirect geological evidence, taken from a wider area, a Lower Devonian age can be assigned with some degree of confidence to the intrusion of the granite. But even when every known adjustment has been made, it will be seen that some age-determinations lie well away from the line adopted as of best fit.

THE HOLOCENE AND THE HOLOCENE-PLEISTOCENE BOUNDARY

We have already mentioned (p. 41) De Geer's chronology of the final stages of the retreat of the ice-sheets across

Scandinavia, based on the counting of varves and hence the
dating in years of the successive moraines. As the ice-sheets
waned, the areas uncovered became colonized by plants. The
spores and pollen of these plants fell in part into bodies of
water and were preserved, to build up a record of successive
plant communities in the same way as the archaeologist
digging trenches into the depressions around an abandoned
city finds in the different layers a sequence of the coins, tools,
etc., of its former inhabitants. Palynology, the study of
spores and pollen, has proved of great value in elucidating
the stratigraphy of the Holocene and the Pleistocene.

The successive layers of the Holocene deposits tell the
story of a sequence of plant communities, clearly reflecting
changes in climatic conditions. This sequence can be traced
from the British Isles across the Low Countries and north
Germany into Scandinavia, whilst a closely similar sequence
has been found around the Alps. The pollen zones shown in

Table 2. *Holocene Chronology*

C^{14} years B.P. (before present)	Pollen Zones	Vegetation (British Isles)	Climatic Trend
2000	IX	Sub-Atlantic (Alder – birch – oak – elm – ash)	Mild
4000	VIII	Sub-Boreal (Alder – oak – ash – ivy)	Deterioration
6000	VII	Atlantic (Oak – elm – alder – ivy	Climatic Optimum
8000	VI	Boreal (Hazel – birch – pine)	Rapid Amelioration
	V		
10000	IV	Pre-Boreal (Birch)	
	III	Younger Dryas (Park Tundra)	Cold
12000	II	Allerød (Birch)	Milder
	I	Older Dryas (Tundra)	Cold

Table 2 can be related to man's development, for his tools occur in pollen bearing layers in lake deposits or on raised beaches, and can also be dated in terms of years by radio-carbon techniques.

The Atlantic Phase, the Climatic Optimum of 6000–7000 years ago, was the time when much of upland Britain was covered by forests, now represented by great spreads of peat, when there was adequate rainfall throughout the Mediterranean area and the desert lands to the south, when the Baltic Basin and the coastlands of much of Europe were submerged by the 'Neolithic Transgression' and when the Great Lakes of North America finally appeared from be-neath the retreating ice-sheets.

The earlier oscillation, the Allerød, is taken as occurring just below the boundary between the Holocene and Pleisto-cene; not only is it traceable throughout Europe, but an equivalent mild phase, the Two Creeks, between two ad-vances of the ice-sheets, is widely recognizable in the region of the Great Lakes of North America. Many radiocarbon dates from the deposits of the Allerød Oscillation agree at it beginning about 10000 B.C. and ending about 8800 B.C. Deposits showing this oscillation are not found north of the great central Swedish moraine (see Fig. 13), so it seems very probable that the retreat of the ice across southern Scandi-navia took place in Pollen Zone II to be followed by a halt when the central Swedish moraine was built up. According to De Geer's varve chronology, the retreat from the central Swedish moraine began about 8100 B.C.; radiocarbon dating places the Zone III/IV boundary, believed to represent the same event, at 8300 B.C., a remarkably good agreement. This is but one of a number of other similarly close links between the absolute dates based on varve chronology and on radiocarbon dating.

THE PLEISTOCENE AND THE PLEISTOCENE/PLIOCENE BOUNDARY

The Pleistocene glaciations consisted of a number of glacial periods separated by interglacials when climatic conditions

must have been much like the present. Moreover the glaciations were often composite, periods of ice advance being
separated by minor phases of retreat, called interstadials,
when there were brief improvements of climate comparable
to the Allerød Oscillation. Each advance of the ice-sheets
was liable to plough off the unconsolidated deposits laid
down during earlier glacial and interglacial periods, so that
the further one attempts to go back into the Pleistocene, the
less complete is the evidence available. Locally the older
deposits have escaped destruction. Gradually as localities
containing valuable records are being discovered, sometimes
in excavations for new buildings or other temporary exposures, the detailed chronology of the whole of the Pleistocene
Period is being pieced together.

But it is only a relative chronology. Few reliable radiocarbon dates are available for events that took place more
than 50000 years ago, whilst the term 'infinite' means a date
older than 70000 years. Shotton in his recent (1966) study of
the last (Würm) glaciation has shown that from Siberia to
the English Midlands and then across Canada and the
northern U.S.A. there is good evidence for a widespread
glaciation lasting from about 10000 to 30000 years B.P.
This was preceded by an interstadial marked by the formation of river terraces and fossil soils whilst before that there
was an earlier episode of the Würm glaciation, which may
have begun about 70000 years ago.

A time-scale for the Pleistocene has been suggested on
astronomical grounds. The Yugoslav geophysicist, M.
Milankovitch, has calculated the varying amounts of heat
that would be received from the Sun, on the Earth's surface
at different latitudes, assuming that the Sun's radiation
remained sensibly constant, but that the position in space of
the Earth relative to the Sun was affected by three factors.
First that the inclination of the Earth's axis to the plane of
its orbit round the Sun, at present $23\frac{1}{2}°$, varied from $24\frac{1}{2}°$ to
$21\frac{1}{2}°$. The periodicity of this obliquity of the orbit is 40,000
years. Secondly, whilst the path of the Earth's orbit round
the Sun was elliptical, the eccentricity of the orbit varied

over a period of 92 000 years. Thirdly, the direction of the Earth's axis in space relative to the 'fixed' stars was not constant, but described a movement like that of an inclined spinning top. This movement, the precession of the equinoxes, had a periodicity of 21 000 years. Combining these three periodicities, Milankovitch obtained by calculation curves of the type illustrated in Fig. 27. His curves show

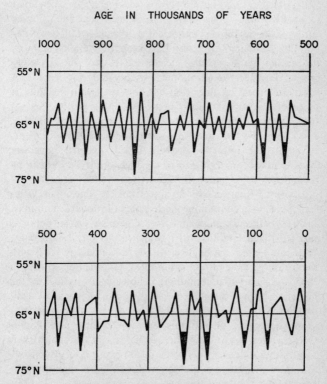

Fig. 27. *Milankovitch's curves showing by the imaginary displacement of latitude, the variation in solar radiation received during the summers of the past million years at latitude 65°N. Periods of cool summers likely to produce glaciation are in black. (After Zeuner.)*

alternations of long periods with predominantly cool or predominantly warm summers. During the longer periods of cool summers snow would accumulate to form glaciers and ice-sheets, but these would melt during the interglacial periods with warm summers. Milankovitch's curves with their built-in time-scale were obtained solely from calculations based on astronomical data. If this time-scale is to be of value for absolute dating, it must obviously agree with the story of climatic variation during the Pleistocene obtained from other lines of evidence. To what extent such curves agree with the purely geological evidence is still very debatable. For one thing, as we have seen, the detailed story of the earlier parts of the Pleistocene is still incompletely known. Further, the astronomical theory does not provide an adequate cause for glaciation nor does it explain the sudden onset of glacial conditions at the beginning of the Pleistocene Period; during the preceding Tertiary Era climatic conditions had been generally warm and humid.

Within the last two decades it has become generally accepted that the beginning of the Pleistocene Period can be recognized by a major change in fauna. In terrestrial deposits this is marked by the appearance of the upper Villafrancian fauna containing the remains of modern mammals such as *Equus, Elephas,* and *Bos*. Where terrestrial deposits can be traced laterally into shallow water marine deposits, there is a significant change at the corresponding level with the disappearance of temperate forms and the appearance of the Calabrian fauna containing cold water forms such as the bivalve *Cyprina islandica*. Recent studies of cores of sediments on the sea bed obtained from depths of hundreds or even thousands of fathoms in the Atlantic and Pacific Oceans have shown that there is a marked change in their foraminiferal content, with the disappearance of discoasters and the appearance of globigerinids at what is thought to be the same level. Above this level the cores show layers yielding an alternation of foraminifera characteristic of warm and of cold water conditions. Unfortunately it is not possible, at present, to correlate with certainty the absolute age of these

Pleistocene fluctuations of temperature in the waters of the oceans with the sequence of glacial and interglacial periods recognizable on the land areas.

Before the recognitions of this world-wide palaeontological break, the base of the Pleistocene in any area was drawn beneath the oldest undoubted glacial deposits to be found. But it is now realized that whilst the beginning of the Pleistocene Period is marked by a world-wide refrigeration of climate, it does not necessarily follow that the refrigeration was everywhere sufficiently severe to produce glaciation. For instance in East Anglia, the Red Crag yielding both a Calabrian marine fauna and containing in its basement bed rolled fragments of Villafrancian vertebrates is overlain by a considerable thickness of Icenian Crags before one comes to the North Sea Drift, an undoubted glacial deposit. The arguments for the inclusion of the Red Crag and the Icenian Crags in the Pleistocene have been strengthened in the last few years by studies of borehole samples from the Icenian Crag. These have shown by both pollen and foraminiferal evidence that the beds contain a sequence of cold and warm faunas and floras. Clearly it was not until well after the beginning of the Pleistocene refrigeration that the ice-sheets had spread from their gathering grounds on the Scandinavian mountains across the North Sea to reach the coasts of East Anglia.

But whilst the position of the Pleistocene/Pliocene boundary can be fixed on fossil evidence, it cannot as yet be precisely dated. For ages around one or two million years only the K/Ar method is available. A number of other techniques are being developed (p. 81), but as yet the ages given by them are not sufficiently mutually consistent. Future work should clarify the situation. At the famous Olduvai gorge, Tanzania, lavas and ashes containing material suitable for K/Ar methods are interbedded with vertebrate-bearing sediments: the available dates from these would place the beginning of the Pleistocene at about 1·75 million years ago, if the African vertebrate faunas can be correlated accurately with those of the European Villafrancian. But certain American geologists working on comparable deposits in the Sierra Nevada suggest an absolute age nearly twice as great. This discrepancy may be due either to differences in the geological criteria by which the boundary has been placed or there may have been at one of the localities loss of radiogenic argon, so that whilst the available dates are all in the correct sequence, they are systematically affected.

We must therefore accept the age of the Pleistocene/Pliocene boundary as lying between 1·5 and 3·5 million years.

THE TERTIARY ERA

This is the best documented part of the Stratigraphical Column. A considerable number of reliable dates, obtained mainly by the K/Ar method, are available either from glauconites in fossiliferous marine sequences or from biotites in volcanic rocks interbedded with vertebrate bearing non-marine beds. As mentioned earlier (p. 78) glauconite ages are somewhat suspect as probably giving only minimum ages, but in the case of many of the Tertiary dates, this is unlikely, for the glauconite-bearing sediments have not been subjected to deep burial or intensive weathering. One difficulty is that whilst independent sequences have been established for the Tertiary beds of Europe and North America, generally acceptable correlation between the beds of the two continents has yet to be made. It is by no means certain, for example, that strata regarded as of Upper Miocene age in Europe are contemporaneous with the so-called Upper Miocene of the American scale. Despite this there are sufficient reliable dates, from the U.S.A., the U.S.S.R., New Zealand and Europe, of horizons near to the boundaries of most of the major divisions of the Tertiary Period, for the dating on the 1964 time-scale to be regarded as substantially accurate.

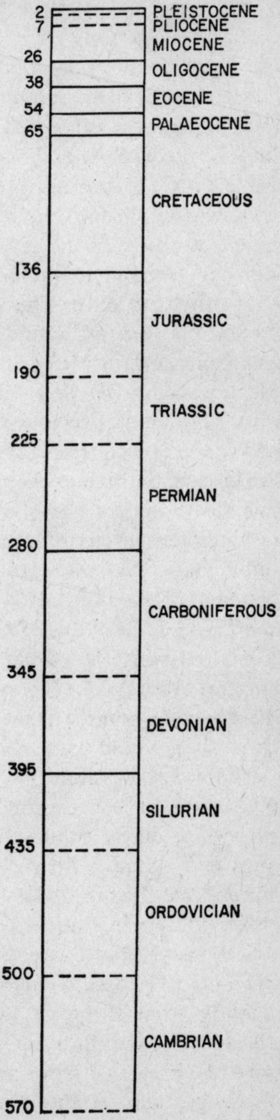

Fig. 28. *The 1964 Time-Scale of the Phanerozoic. Solid Lines – system boundaries dated with precision. Broken Lines – system boundaries imprecisely dated on the available evidence.*

THE MESOZOIC ERA

The Upper Cretaceous and the greater part of the Lower
Cretaceous rocks are reasonably well documented. Dates are
available from fossiliferous glauconite-bearing sediments of
Europe, the U.S.A. and the U.S.S.R. and also from micas
and feldspars in igneous rocks whose stratigraphical position
is known within fairly narrow limits. For the Lower Cretaceous beds most of the available dates are from glauconites.
Many of these give younger dates than would be expected
from other evidence. The age of 136 million years for the
Cretaceous/Jurassic boundary is determined mainly by the
ages from certain rocks of highest Jurassic age. But below
this, the accuracy is very much lower. In the 1964 symposium
a total of merely twelve dates is listed from the whole of the
Jurassic sequence. These are thought to be accurate within $\pm$
20 per cent but five of them were determined on glauconites
and are thought to give only a minimum age. There is an
even greater scarcity of information from the beds of Triassic
and highest Permian age. Table 1 shows that in the 1964
symposium the duration of the Jurassic Period was increased
and that of the Triassic Period substantially decreased as
compared with the earlier estimates of Holmes and Kulp,
but in the text it is freely admitted that the evidence on which
these system boundaries were dated is by no means as precise as one would wish.

The richly fossiliferous rocks of the Cretaceous System
in Europe have been subdivided into twelve stages. These
stages are major units, used in the detailed correlation of
beds of England with those of Germany, Russia, Southern
France, etc. Each stage is built up of a number of zones, the
majority of which are of limited geographical extent. The
stages have been erected entirely on palaeontological
grounds, the fossil assemblage of each stage differing significantly from those of the stages above and below. The
thickness of the different stages varies considerably, both in
any one area and from area to area. On strictly geological
evidence there is therefore no means of determining the

period of time (the Age) represented by the beds of each stage. If we make the arbitrary assumption that the ages of the Cretaceous are all of equal length, then dividing the duration of the Cretaceous Period, 72 million years, by 12, we obtain the figure of 6 million years for each age. The relatively few reliable dates, mainly from the Upper Cretaceous beds, that can be placed stratigraphically within a particular stage, fall fairly well into their place on this arbitrary scale. Many more dates are needed, particularly from the Lower Cretaceous beds, before it will be possible to erect a time-scale for the Cretaceous Period showing if there were significant variations in the duration of the different age divisions.

It is impossible to go as far as this with the Jurassic strata. This is particularly unfortunate, for many of the principles for the palaeontological subdivision and correlation of beds were hammered out on the rocks of the ten stages of the European Jurassic. Each of the Jurassic stages has been subdivided into a number of zones. A total of more than forty zones is generally accepted, so that taking the 1964 figures of 50–55 million years as the duration of the Jurassic, one can regard each zone as spanning approximately a million years.

THE PALAEOZOIC ERA

The base of the Permian and the base of the Devonian are the only two system boundaries that can, at present, be regarded as at all closely dated. It is particularly unfortunate that so few reliable dates are available for the Silurian, the Ordovician and the Cambrian, the three systems that make up the Lower Palaeozoic, and especially that the age of the base of the Cambrian is so uncertain. This means that the first 200 million years of Phanerozoic time are dated, but only approximately. It is interesting to note that the 75 million years or so of the Ordovician Period span five ages and twelve graptolite zones, giving to a first approximation 15 million years to an age and 6 million years to a graptolite zone, very different figures from those based on ammonite zoning for the Cretaceous.

Towards the close of the Devonian Period the ammonoids were present in sufficient variety and abundance to be used for the definining of zones. If we take 50 million years as the duration of the Devonian Period, then each of the six ages shown on Fig. 26 can be regarded as spanning about 8 million years. There are nine ammonoid zones in the highest stage, so that, as for the Jurassic, we have a figure of about one million years for each zone. But Fig. 26 also shows the very considerable differences in the maximum known thickness of the beds of each stage. This underlines the uncertainty in regarding the stages as of equal duration.

The Carboniferous rocks present many problems. In the first place there are not only marked contrasts in the lithology of the Carboniferous rocks of north-west Europe as compared with those of the U.S.A. and the U.S.S.R., and very different rocks laid down during the Permo-Carboniferous glaciation of the Southern Hemisphere, but also the beds of north-west Europe show unusually rapid variation vertically and horizontally, both in their mass characters and their fossil content. As a result the detailed correlation of many parts of the succession of the Carboniferous rocks throughout Europe is as yet still undecided. Whilst figures varying between 10 and 20 million years were put forward in 1964 for the duration of the five major divisions of the Carboniferous Period, these were based on slender evidence and it is at present impossible to go further. The position as regards the stages of the Permian System is unfortunately no better.

MOUNTAIN BUILDING MOVEMENTS

On p. 21 we referred to the numerous orogenies which have affected the Phanerozoic rocks and suggested that each of these mountain building episodes had a complex history. A sequence of events, involving structural deformation, metamorphism and the emplacement of intrusive bodies of igneous rocks, can be worked out by normal geological methods, but the various stages on this relative time-scale cannot usually be tied into the Stratigraphical Column with certainty. The development of isotopic dating promised an entirely new line of attack, but up to now the progress has not been as great as was at first anticipated.

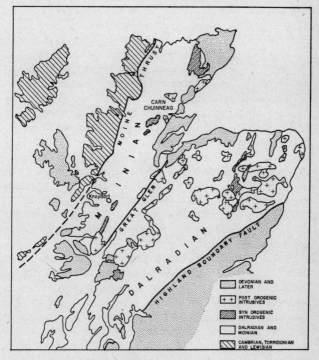

Fig. 29. *Simplified Geological map of the Scottish High lands.*

The rocks of the Scottish Highlands, between the Moine Thrust and the Highland Boundary Fault (Fig. 29), have been investigated in greater detail from the geological, structural and isotopic angles than any other orogenic belt in the world. A discussion of this area will show the present difficulties in interpreting the significance of the numerous isotopic dates that are available.

On the foreland to the west of the Moine Thrust fossiliferous Cambrian rocks rest unconformably on the Pre-Cambrian (Fig. 8). To the east of the Thrust and extending southwards and eastwards to beyond the Great Glen, outcrop a rather monotonous series of metamorphosed arenaceous sediments, the Moinian. These rocks are overlain by the lithologically more varied Dalradian. Both the Moinian and the Dalradian are complexly folded, metamorphosed and are intruded by igneous rocks. Unfolded rocks of Devonian age rest unconformably on the Moinian, the Dalradian and their intrusive rocks (see Fig. 29), therefore the Caledonian orogenic movements must have ended before these Devonian (the Old Red Sandstone) rocks were laid down. On palaeontological evidence, the Old Red Sandstone of the Highlands is certainly of Middle Devonian and possibly the basal beds of Lower Devonian age. The discordant Newer Granites, which cut the folded Moinian and Dalradian rocks, have yielded isotopic ages of about 395 million years at a number of localities. On the Phanerozoic time-scale (Fig. 28) this is very close to the Devonian/Silurian boundary.

During the last two decades, detailed structural analysis of the Dalradian rocks has shown that they have been affected by several episodes of folding. The first folds are huge recumbent structures (see Fig. 30), one of which, the Ben Lui syncline, has been traced for 175 miles from Kintyre to Deeside. These recumbent folds strike south-west to north-east. They have been folded by crossfolds with a general northwesterly–southeasterly strike, which are best developed in the Central Highlands. The third folding phase produced sharp asymmetrical folds with north-east–south-west or east–west strike. A number of metamorphic episodes have also been recognized. During the first phase the clay rocks were converted to silvery micaceous phyllites. During the second phase there was first some development of garnet and biotite and, after the intrusion of dolerites, of higher grade metamorphic minerals such as andalusite, kyanite, staurolite, etc. This was the climax of metamorphism and seems to have been associated with the emplacement of the syntectonic (Older) igneous bodies. Then followed retrogressive metamorphism shown by the chloritization and mechanical breakdown of the earlier formed metamorphic minerals.

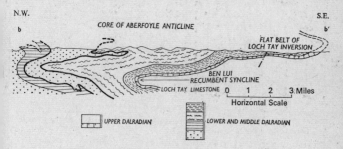

Fig. 30. *Section showing the geometry of the great recumbent folds of the first folding episode. (By permission from* The Geology of Scotland, *by M. W. R. Johnstone, Oliver and Boyd, 1965, Fig. 4.12b, p. 133.)*

The time relationship between fold movements and metamorphism can be shown by another line of evidence. The minerals produced as a result of metamorphism are dependent on the conditions of temperature and pressure which have affected the rocks. The low-grade metamorphic minerals (chlorite and biotite) were produced under less intense conditions than high-grade minerals such as kyanite, andalusite and sillimanite. Therefore, by mapping the distribution of the various metamorphic minerals one can obtain a picture showing the different intensities of metamorphism. This has been done for parts of the Central Highlands and, as is shown by Fig. 31, the distribution of the metamorphic zones is a relatively simple one, contrasting markedly with the structural complexities of the same area. The metamorphic grades clearly cut across and are therefore later than the last folds.

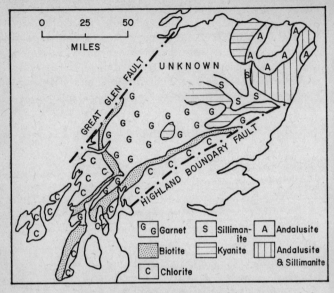

Fig. 31. *The Metamorphic Zones of the Grampians. (After Johnstone, 1966.)*

The sequences of structural and metamorphic events recognizable in the Dalradian rocks is tabulated below:

	Emplacement of Younger Granites
F_3 on N.E.–S.W. or E.–W. planes	M_4 Retrogressive metamorphism
	M_3 Maximum metamorphism Emplacement of Older Granites, etc.
F_2 crossfolds	M_2 Development of garnet and biotite
F_1 major Recumbent folds	M_1 Clay rocks – phyllites

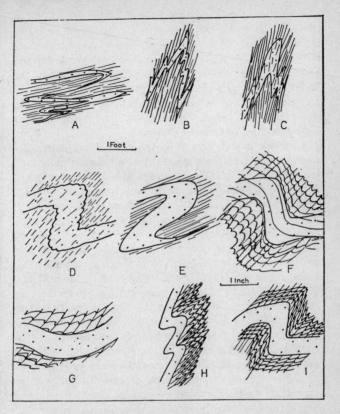

Fig. 32. *Varieties of folds in the Dalradian of Scotland:*

A, B, C – *diagrammatic styles of* F_1 *folds (near Kinloch Rannoch, Sron Mhor and Aberfoyle respectively)*

D, E, F – *diagrammatic styles of* F_2 *folds (near Kinloch Rannoch)*

G, H, I – *diagrammatic styles of* F_3 *folds (near Kinloch Rannoch, Ben Lui and Pitlochry respectively)*

Scale in feet refers to A, B, C, D *and* E; *scale in inches refers to* F, G, H *and* I.

(By permission from The British Caledonides, *Oliver and Boyd, 1963; N. Rast, Fig. 2, p. 127.)*

A number of fold episodes have also been recognized in the Moinian rocks, but owing to their relatively uniform lithology, it has not so far been possible to disentangle their metamorphic history as fully as has been done for the Dalradian. The west-north-westerly movement along the Moine Thrust, with a horizontal displacement of at least 10 miles (though 75 miles has been suggested), must have taken place later than the maximum of metamorphism and may have continued into the Devonian.

A considerable number of K/Ar and Rb/Sr age-determinations have been made on micas from both the Dalradian and the Moinian rocks. Those from the Moinian show a spread from 400 to 430 million years with a peak around 420 million years, whilst the Dalradian dates range from 430 to 490 million years with a peak around 440 million years. It is the significance of these dates that is under debate. One view is that the 420 date from the Moinian is that of the main metamorphism, the somewhat older ages from the Dalradian being due to incomplete overprinting of an earlier metamorphism, all trace of which has been destroyed in the Moines. Another interpretation is that the Moinian and the Dalradian dates are not related directly to metamorphism, but that they record the stage in the post-metamorphic uplift (p. 36) and cooling history of the mountain belt when the atoms of radiogenic daughter nuclides were no longer able to diffuse out of their host rock or mineral, and were retained. As the Dalradian rocks were less deeply buried, they would cool the more quickly and therefore record the older age. In this connection it is interesting to note that analyses of muscovites and biotites made from the same specimens of Dalradian rocks record significantly older dates ($+ 12$ million years) from the muscovites than from the biotites. Moreover, recent determinations of whole-rock K/Ar ages from slates in high structural positions within the fold belt give consistently older ages, up to 490 million years, than those obtained from the specimens collected from lower structural levels. The oldest of these 'high level' dates are regarded as giving a close approximation to the time of the main metamorphism, the younger dates as reflecting the delay before the more deeply buried layers passed through the threshold temperature of argon retention. As we have already mentioned (p. 76) the regionally metamorphosed Carn Chuinneag intrusion has yielded a Rb/Sr whole isochron of 530 million years and similar ages have been obtained from several other geologically comparable intrusions. On this view, therefore, the age of the main metamorphism is between 490 million and 530 million years, that is, in terms of the Phanerozoic time-scale (Fig. 28) it occurred towards the end of the Cambrian Period, instead of in late Silurian times as would be the case if the 420 'event' is accepted at its face value.

In Glenelg and Morar the Moinian rocks are considerably less altered and tectonically deformed than in the other parts of the Highlands. The effects of the last fold phases have been less intense and therefore the earlier phases have been not obliterated by overprinting. In these areas the Moinian has yielded K/Ar dates of as far back as 562 million years, whilst a Rb/Sr age of 740 million years has been obtained from a pegmatite in Knoydart, which cuts and is therefore younger than the Moinian country rocks.

Whilst the precise chronology of the Caledonian mountain belt has yet to be elucidated, it is apparent that its history is both more complicated and also spans a considerably longer time-period than was thought to be the case only a few years ago. It is also a warning against taking too simple a view of the chronology of other mountain belts, especially when they have not been investigated in detail or only a few isotopic dates are available.

PRE-CAMBRIAN TIME

The greatest contribution to geological knowledge made by the recent advances in isotopic dating is in proving the vast duration of time that is represented by the Pre-Cambrian strata. As yet the coverage of dates from the Pre-Cambrian is somewhat meagre and is concentrated in a limited number of areas. But in these areas, the geochronology of the Pre-Cambrian is being established. In time, the correlation, even the intercontinental correlation, of the events of Pre-Cambrian times will be established on a firm basis. Isotopic dating seems to be the only method by which this can be done, for whilst the rocks of Pre-Cambrian age are not completely devoid of organic remains, these have only been found at widely scattered localities. It seems extremely unlikely that even in younger Pre-Cambrian rocks sufficiently abundant fossil evidence will ever be found for it to be used, as in the Phanerozoic, for purposes of correlation.

One of the most thoroughly studied areas of Pre-Cambrian rocks extends along the west coast of Scotland to the west of the Moine Thrust. As already mentioned on p. 28, a group of unmetamorphosed sediments, the Torridonian, rests unconformably on the greatly metamorphosed Lewisian Complex. The rocks of the complex had been subjected to two orogenies, the Laxfordian and the Scourian. Twenty years ago it had been argued on purely geological grounds that the two orogenies must have been separated by a long time gap: this has been confirmed by isotopic dating. Numerous dates of between 2200 and 2600 million years have been obtained from the Scourian and of 1200–1600 million years from the Laxfordian. The suite of dolerite dykes, which were intruded after the Laxfordian metamorphism, give dates around 2200 million years. The sediments and igneous rocks affected by the Scourian orogeny must be older than 2600 million years, but how much older we cannot tell. It is generally agreed on geological grounds that the Torridonian or parts of it are represented to the east of the Moine Thrust (Fig. 29) by some of the Moinian rocks. The Moinian rocks must be older than the Knoydart pegmatite, which cuts them and has a Rb/Sr age of 740 million years, but how much older is not yet known. Recent K/Ar and Rb/Sr determinations on detrital feldspars, micas and pebbles of acid igneous rocks from the Torridonian have given ages in the range of 1700–1150 million years. These relate to the rocks from which the Torridonian sediments were derived, so the Torridonian must be younger. Whole-rock Rb/Sr determinations on shales from the upper beds of the Torridonian of (751 million years) and for the lower beds (935 million years), show that the Torridonian was deposited during an unexpectedly long time-span. The available isotopic data is not sufficiently precise to either prove or disprove the Torridonian/Moinian correlation suggested on purely geological grounds. This is unfortunately typical of the present accuracy of dating the Pre-Cambrian rocks or the events represented by them.

The older geologists were greatly impressed by the fact that in many areas of Pre-Cambrian rocks, one could recognize, as in western Scotland, a younger group of unmetamorphosed or but slightly metamorphosed rocks resting unconformably on strongly metamorphosed strata. The younger unmetamorphosed beds were often referred to as the Algonkian, the underlying more intensely metamorphosed rocks as the Archaean. It was thought that these represented two major divisions of Pre-Cambrian times, but isotopic dating has shown that mere appearance of the rocks can be very misleading as regards true age in Pre-Cambrian terrain.

For example, in Canada an area of greatly metamorphosed rocks, unusually rich in marbles for the Pre-Cambrian, extends from southern Labrador to the Great Lakes. These rocks, the Grenville Series, were recognized over 100 years ago and from their 'old' appearance were regarded as Archaean and, therefore, as of very considerable antiquity. But they have yielded numerous isotopic dates of around 1000 million years, proving that they owe their 'Archaean' appearance to having been affected by a major orogeny with much plutonism in late Pre-Cambrian times. In parts of the South African Shield on the other hand, there are great thicknesses of but slightly deformed and virtually unmetamorphosed sediments and volcanic rocks of an 'Algonkian' aspect. Yet not only are they intruded by the Bushveld Igneous Complex, with a mean age of 1940 million years, but even older ages of 2000 to 2200 million years have been obtained from monazites and uraninites in the Main Reef, the gold-bearing conglomerate of Johannesburg.

It is, therefore, apparent that the older ideas of the relative dating of the Pre-Cambrian rocks by their appearance can be quite misleading. In the Grenville Province the rocks owe their 'Archaean' aspect to the effects of a late 'Algonkian' orogeny; in South Africa, on the other hand, 'Algonkian' looking sediments are of very considerable antiquity, indeed are older than much of the Lewisian ('Archaean') Complex of western Scotland.

Isotopic dating is enabling us to build up a picture of the successive Pre-Cambrian orogenies and to delimit the areas affected by each. Sufficient dates are now becoming available from the United States and Canada, both from surface outcrops and from the numerous deep boreholes which have

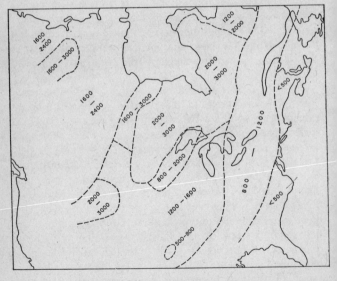

Fig. 33. *Age Zones in the Pre-Cambrian rocks of the Canadian Shield and adjacent areas. (Based on* Nuclear Science Series Report No. 41, 1965.)

penetrated through the Phanerozoic rocks that cover the underlying Pre-Cambrian, for it to be possible to map in broad outline at least the age zones of the Pre-Cambrian (Fig. 33). Broad dating of the fold belts recognizable in the rocks of the Baltic Shield is also possible (Fig. 34). Certain major geological events can now be traced over very considerable distances. For example, a great suite of basic dykes emplaced between 2200 and 2300 million years ago can be recognized from north-west Scotland through the Pre-Cambrian rocks of Greenland, the Ungava Peninsula in Quebec to the north-west territories of Canada.

For the earlier part of the Pre-Cambrian times, with ages greater than 3000 million years, only the vaguest outlines of

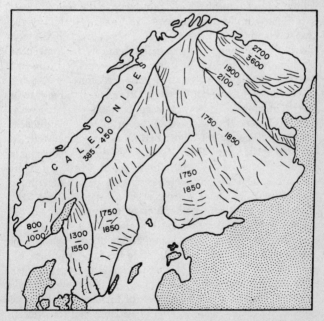

Fig. 34. *The Isotopic Ages of the Orogenic Belts recognizable in Scandinavia. Stipples – areas of undeformed Phanerozoic strata (From various sources.)*

the picture are, at present, available. In certain areas, such as
Rhodesia and parts of the U.S.S.R. (the Ukraine and the
Kola Peninsula) where the basements rocks have not been
overprinted by later events, isotopic dates ranging back to
nearly 3500 million years have been obtained.

HISTOGRAMS

Another way of handling isotopic data is the construction of
histograms such as Fig. 35. Only a minority of the age-
determinations for the Phanerozoic rocks are from sediments, those from the Pre-Cambrian rocks are almost
entirely from metamorphic and igneous rocks. Such histograms have therefore been held to record the waxing and
waning of metamorphism and plutonism in the area covered.

It is very important, however, to consider carefully the
data used in the construction of a histogram. In the first
place, the sampling for age-determinations has not been
random, but has been concentrated in areas, such as the
Scottish Highlands which have been the centre of research
activity and of much controversy. Other areas, thought to be
less problematic, have not been covered so intensively.
Therefore, certain peaks may be artificially heightened;
similarly, some of the troughs, periods of diminished activity,
may be misleading. For instance, in Fig. 35A, there is a
distinct trough around 600 million years. Since that figure
was constructed, a considerable number of age-determinations have been made on the rocks of the Channel Isles.
These give very consistent dates of between 580 and 590
million years for a significant igneous event which affected
the Channel Isles, whilst comparable figures have been
obtained for an event which affected the Pre-Cambrian, but
not the Cambrian, rocks of the Malvern Hills, Anglesey, etc.
The recognition of this 'Cadomian' episode may well have
repercussions on the age adopted for the base of the Phanerozoic, for there is clear geological evidence that fossiliferous
sediments of Cambrian age rest on an erosion surface cut
across plutonic rocks yielding Cadomian dates. The magni-

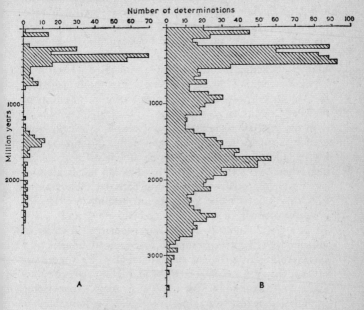

Fig. 35.

A. *The number of published age-determinations for metamorphic migmatitic and granitic rocks from the British
Isles plotted as a histogram against time. Dates obtained
by all methods are included.*

B. *A similar histogram for comparison showing the
distribution of 1600 dates from all parts of the world,
published during the years 1960–2.*

(Reproduced with permission from Qtrly. J. geol. Soc.
Lond. 1965, **121**, *p. 480.)*

tude of the time gap between this latest Pre-Cambrian event
and the deposition of the Cambrian sediments is a problem
for the future.

Another possibility to be considered is that a marked
trough or even an absence of dates in any one area may be

the result of complete overprinting and obliteration by a later metamorphic event. For instance, the zone between 800 and 1150 million years is blank on the histogram for the British Isles, yet is represented on the world histogram by a marked peak corresponding to the Grenville Orogeny in Canada and the U.S.A., and the Telemark Episode in southern Scandinavia. It is possible that this fold belt may have extended through the northern part of the British Isles, but that its relics are unrecognizable owing to the effects of the later Caledonian movements.

In the past all available isotopic dates have usually been used in the construction of histograms. But this may be misleading, for some of the dates may be cooling dates (p. 104) or overprinted dates (p. 104) and therefore do not relate to a significant geological event. As Moorbath (see bibliography) says (p. 123) 'For the purposes of defining episodes of crystallization in a basement complex, a single U/Pb, Rb/Sr or K/Ar date on a really resistant (i.e. in terms of diffusion of radiogenic isotope) coarse-grained mineral, or a single Rb/Sr whole-rock isochron date, may be far more significant than any number of K/Ar and Rb/Sr dates on fine-grained micas. Though the latter may form a convincing peak on a date histogram, this may signify no more than the termination of a radiogenic Ar or Sr diffusion episode. In contrast, a considerable spread of dates on a histogram may often be due, not to continuous plutonic activity, but to the inclusion of a large number of apparent "overprinted" dates – especially in a poly-metamorphic terrain.'

LIFE DURING THE PRE-CAMBRIAN

At the beginning of Phanerozoic times, certain groups of invertebrates began to secrete hard parts and have therefore been preserved as fossils. The trilobites (Fig. 15) of the Lower Cambrian beds are complex and specialized organisms. They must have had a long ancestry, but as it was a soft-bodied ancestry no definite indications of it have so far been discovered. The development of hard parts in organisms was a major event in Earth-history, but we do not know precisely when this occurred or the reasons for it. Nor do we know when or how life originated on the Earth's surface. But, thanks to isotopic dating, we are now certain that it must have occurred thousands of millions of years ago.

The Pre-Cambrian rocks of many parts of Africa, the U.S.S.R., the U.S.A., China and Australia contain stromatolites, carbonate bodies with a regular laminated structure, similar to those formed in the Phanerozoic rocks by certain lime-secreting blue-green algae. It is often very difficult to draw the line between true stromatolites, especially when not too well preserved, and various other bodies of rather similar appearance but of inorganic origin. But the presence of undoubted stromatolites of algal origin in the Pre-Cambrian rocks at various horizons and in several continents is now generally accepted. Indeed the Russians claim that in certain of their younger Pre-Cambrian rocks stromatolites of distinctive form are restricted in their vertical distribution, but as the time-range of the different assemblages seems to be measurable in terms of hundreds of millions of years, they do not provide a very precise means of correlation. Stromatolites range back to, at least, 2500 millions years ago. Indeed their occurrence has been claimed in rocks dated as 3500 million years old in the Fig Tree Series of the Swaziland System.

In addition, cherts from the Gunflint Formation of the Lake Superior region, have yielded traces of the filaments of algae and possibly of fungi and bacteria. These rocks are dated at about 1900 million years by the K/Ar method. Comparable structures have very recently been reported from rocks of about the same or even older age from Guyana and Western Australia.

As well as these remains of algae, considerable number of problematica have been described from the Pre-Cambrian; problematica in the sense that not only is their organic origin debatable, but that they may belong to phyla that are not now living or even represented in the fauna of the Proterozoic rocks.

THE AGE OF THE EARTH'S CRUST

Meteorites are believed to be debris from a planet, which was shattered by some cosmic catastrophe at about the same time that our planet originated. They provide a means of going back beyond the oldest rocks that have so far been found in the Earth's crust. The Older Granite of Swaziland with an isotopic age of 3400 million years or more must have been intruded into pre-existing rocks. We are therefore trying to probe back to the time when the Earth's crust solidified.

Meteorites are of two major kinds – *iron meteorites* composed mainly of nickel-iron alloy, and *stony meteorites* consisting largely of silicate minerals. The internal structures of the iron meteorites indicate that they were formed under conditions of high temperature and pressure, but the make-up of the stony meteorites is more varied, with material that formed under high pressure mixed with silicate minerals which crystallized under conditions of lower pressure and more rapid cooling.

Although the concentrations of the requisite elements are very low, the age of certain stony meteorites has been determined by the Rb/Sr or by the K/Ar methods. Consistent ages of 4500–4600 million years have been obtained. Similar dates have been given by the measurement of the isotopic constitution of the very small amounts of radiogenic lead present in both stony and iron meteorites.

There is therefore a period of at least 1000 million years to be accounted for between the crystallization of the oldest known rocks on the Earth's surface and the consolidation of the Earth's crust (Fig. 36).

THE AGE OF THE MOON'S CRUST

The samples of lunar rocks and dust brought back by the Apollo 11 and 12 missions have been investigated with the utmost thoroughness. Age-determinations have been made using the Rb/Sr, K/Ar and U/Th/Pb methods. The Apollo 11

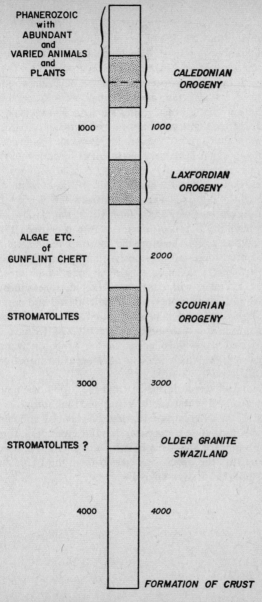

Fig. 36. *The Isotopic Time-Scale from the formation of the Earth's crust.*

specimens from the Sea of Tranquillity consisted of almost equal amounts of crystalline igneous rocks of a basaltic composition and of lunar soil and microbreccias. The soil was made up of fragments of crystalline rocks, glass and iron meteorites, whilst the microbreccias were a compacted mechanical mixture of soil and small rock fragments. The Apollo 12 mission landed not in the centre of a marie, but very close to the rim of a crater in the Ocean of Storms. The samples brought back were nearly all of igneous rocks with very few microbreccias.

The results of the age-determinations are consistent and extremely interesting. The basalts from the Sea of Tranquillity gave ages of 3700 million years, the closely similar rocks from the Ocean of Storms, a spread between 1700 to 2700 million years. Therefore the marie are of different ages, as had been suggested previously from the interpretation of the photographs taken on earlier missions, which showed a pattern of craters with the rims of the older ones cut by the unbreached younger ones. The samples of soil and of microbreccia gave older ages of 4600 million years. Some of the fragments in the microbreccias are believed to have been transported from outside the marie, probably from the Highland areas which were to have been the target of the dramatic Apollo 13 mission.

These ages therefore suggest that the Moon originated at the same time as the Earth, about 4600 million years ago. Certainly the differences in the detailed chemical and mineralogical composition between the basaltic type rocks from the Moon's surface and terrestrial basalts do not support the theories that the Moon was ejected from the Earth during Pre-Cambrian or even later times.

7. Conclusions

The science of geology brings together the physico-chemical and the biological sciences. The study of the composition of the rocks and minerals of the outer layers of the Earth's crust is essentially a branch of inorganic chemistry. The ways in which these substances were formed are largely deduced by applying physico-chemical and physical principles to interpret the features observable primarily in the field but also in the laboratory. The nature of the deeper layers that make up the Earth cannot, however, be observed directly, but have to be inferred by studying the manner in which the Earth reacts to physical phenomena, such as the passage of waves generated by earthquakes or man-made shocks. Many of the rocks in the Earth's outermost layer contain fossils. The vertical and horizontal distribution of these fossils can be determined by observational studies of surface exposures and the cores from boreholes. To interpret the conditions under which they may have lived we must draw on the knowledge of biologists of the nature and factors that control present-day animal and plant distribution.

The geologist has to invoke another dimension, that of time. Time is only exceptionally of importance to the modes of thought of the chemist, the physicist and the biologist. But in most branches of geology, one is constantly having to consider the changes that may have occurred during the passage of time, whether it be in the study of the major and minor topographic forms of the land masses or ocean basins, or the investigation of the sequence of the stratified rocks and their contained fossils, or in the nature and origin of mineral deposits and hence the places, as yet undiscovered, where they may be found, etc., etc. The geologist has to take into account the progression of changes that occur in rock material, first as a result of its burial beneath the weight of later formed deposits, then by contact with molten or potentially molten magma or by compression during mountain building movements, and finally, as these rocks are uplifted and exposed on the Earth's surface, the attack of the weathering agents, whose effects will be partly determined by the length of time during which they are able to act.

In the earlier chapters we have outlined the gradual appreciation of the importance of time in the explanation of geological phenomena, then the building up of the Stratigraphical Table, the relative time-scale, and the attempts to convert the relative time-scale into an absolute time-scale with a million years as its unit. We have traced the development of the present isotopic time-scale by the application of extremely specialized chemical and physical techniques. But the 'absolute' ages obtained by isotopic methods must pass the acid test of placing the rocks and minerals that have been 'dated' in the same relative order as that in which they have been proved to occur by investigations using standard geological methods. If they fall into a different order, then the 'absolute' ages must be suspect.

During the past two decades there has been a great burst of activity, including the development of new techniques of isotopic dating, which are applicable to a much greater variety of rocks and minerals than were the pioneer ones. But, as we have seen, many of the major units of the Phanerozoic part of the Stratigraphical Table are by no means precisely dated. A great deal remains to be done before we have a world-wide picture of the sequence of even the major events of Pre-Cambrian times, or a reasonably precise knowledge of the duration and chronology of the mountain building episodes of post-Pre-Cambrian, let alone of Pre-Cambrian times.

As more and more isotopic dates become available on rocks and minerals whose stratigraphical position is known within narrow limits, as the present techniques are refined and as new ones are developed that can be applied to materials at present undatable, so the isotopic time-scale will become both increasingly precise and also more comprehensive.

Let us now consider the effects of this, one of the major growing points of geology, in giving a precision, at present unobtainable, to many branches of geological investigation and hence to human knowledge.

THE PHANEROZOIC TIME-SCALE

As we have seen (p. 95), only the last 100 million years or so of the Phanerozoic time-scale can be regarded at present as being dated with a reasonable degree of precision. Further back in time than this, the number of isotopically and stratigraphically fixed points decreases markedly and, in particular, the absolute dating of the first 200 million years of the Phanerozoic can at best be regarded as but approximate. Moreover, whilst the available evidence suggests that the ages into which periods are divided may have had a duration of six to eight million years, we cannot usually be sure about which ages of a period were significantly longer or shorter than others. As regards the smaller units, the zones, the available evidence suggests that those based on ammonoids, in the rocks of the Devonian and Jurassic Systems, represent a time-span of the order of a million years, many times shorter than that of the graptolite zones used in certain of the Lower Palaeozoic rocks.

By recording on maps the distribution of the various lithologies shown by beds believed to belong to the same stage, or, for smaller and more thoroughly known areas, the same zone, we can construct palaeogeographical maps which show the extent of the several environments inferred to have been present during that particular time interval of the geological past. But we cannot be sure about the length of the time-span covered nor that all the beds in question are really the time-equivalents of each other.

It seems completely utopian to hope that isotopic dating will become so precise and, perhaps more important, capable of being applied to a so much wider variety of rocks and minerals than at present, that it will supplant the present palaeontological methods of correlation. But if we could produce a comprehensive Phanerozoic time-scale, based on points whose isotopic age and stratigraphical position were known with sufficient precision for the great majority of the system and stage boundaries to be fixed with an accuracy of even two to three million years, then our knowledge of Earth-history during the Phanerozoic would gain greatly in definition and precision.

THE RATE OF EVOLUTIONARY CHANGE

On p. 48 we gave a summary account of the development of animals and plants during the Phanerozoic to show the course that organic evolution has taken. The tempo of evolution must also be considered; this has differed not only from one group of organisms to another but also within the members of any one group. For example, amongst the cephalopods, the nautiloids of the Lower Palaeozoic show considerable evolutionary change, but as this occurs at a relatively slow rate they are of limited value in the correlation of beds, for each nautiloid genus spanned a considerable time range. In Upper Palaeozoic, and particularly in post-Palaeozoic times, the nautiloids have changed but little. Indeed the living *Nautilus* is almost identical with specimens to be found in strata of Jurassic age. From the nautiloid stock, developed in the Devonian period, first the goniatites and then from certain of these the ammonites of the Mesozoic (Fig. 15); both the goniatites and the ammonites (especially the latter), show very rapid evolutionary changes, so that they are of great value in the zonal subdivision of the beds and, therefore, in correlation. The ammonites became extinct by the close of the Mesozoic, but the unprogressive nautiloids have survived to the present.

One can show, as in Figs. 15 and 16, the changing importance of groups of organisms with time. In each case, there is the initial phase when new families, genera and species are appearing, until the group reaches its acme, with maximum variation, abundance and distribution. Then follows the period of decline, which may be quite short or prolonged. During this phase new genera and species still appear, but they are outnumbered by the families, genera and species that die out, until the group itself passes into extinction.

But it is a much more difficult matter to present a truly quantified picture of the variation in the rate of evolutionary change. The number of taxonomic units (families, genera or species) of a particular group known from a certain horizon is determined to a considerable extent by the degree of detail in which the group has been studied. The criteria used for the recognition of taxonomic units in fossils cannot be other than subjective. The intensive study of a hitherto long neglected group of fossils will almost inevitably mean that the old 'broad' genera and species have been split up into a large number of new and more precisely defined units. Graphs showing rate of evolution in terms of the number of orders, families, genera or species recorded from the successive systems or, as in Fig. 37, the number of new genera recorded, reflect to a considerable degree the varying activities of systematic palaeontologists: like histograms (p. 110), they require interpretation. In this case considerable knowledge of the history of research into the group in question is necessary to assess the probable evolutionary significance of the various maxima and minima indicated.

Another method, used in Fig. 38, is a 'profit and loss' account, based on the difference across each stratigraphical horizon considered of the number of classificatory units or taxa appearing for the first time and the number of taxa appearing for the last time. This figure is based on data in *The Fossil Record*, a symposium published by the Geological Society in 1967. This is a compilative work in which numerous specialists have listed the first and last recorded occurrence of 2526 taxa in terms of 72 stratigraphical horizons, and on a world-wide basis. It is the definitive work on the time range of fossils down to the size of the taxa considered. Unfortunately, these vary considerably in size, for each

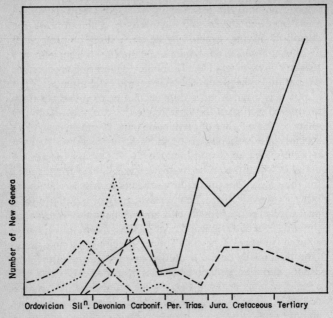

Fig. 37. *Rate of appearances of new genera in the four groups of fish (after Simpson). Firm Line – Bony fish; Sharke like fish; Dotted line – Broken Line –Placoderms; Dot-Dashed Line – Agnathids.*

contributer could be allowed only a limited amount of space and, therefore, those dealing with large and varied groups such as the *Ammonoidea* had to use much larger units for their taxa than those concerned with much smaller groups, which in some cases could be treated down to generic level.

Bearing in mind the points already made as to the uncertainties introduced by differing degrees of palaeontological knowledge, these curves give a broad picture of the evolutionary change with time. First there was rapid evolution, amongst the invertebrates in Cambrian and early Ordovician times. This was followed by a long drawn out but gradual waning culminating in the extinction during the Permian of many taxa that were characteristic of the

Palaeozoic faunas and floras, in all groups, except the vertebrates. In the later part of the Triassic Period there was another burst of evolutionary activity, delayed in the case of the vertebrates, to be followed by fluctuations ending in a general waning during the later part of the Tertiary Era.

But this is considering evolutionary change very much in the large. The fossil record also includes a number of evolutionary series in which we can trace in detail the development of a species from an earlier one. This can be

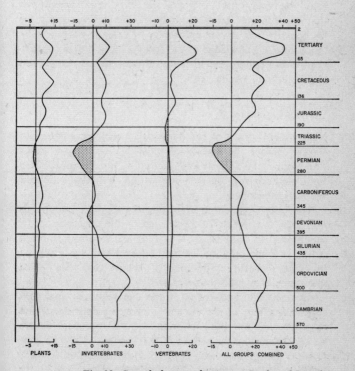

Fig. 38. *Smoothed curves showing net Profit and Loss for the chief groups of organisms.*
(Based with permiss ion on Figs. 19, 20, 21 and 22 of The Fossil Record, 1967, *Geol. Soc. Lond., and* Trans. geol. Soc., Glasgow, 1947, **21**, *Fig. 3.)*

done, for example, in the case of certain ammonites found in the Upper Jurassic rocks of England, of certain trilobites from the Upper Cambrian beds of Sweden, and perhaps best known, the heart urchins which lived in the soft oozes on the floor of the sea in which the Chalk was deposited. The sequence of fossil horses collected from the Tertiary rocks of the Western United States provides a longer and more complex story and the rate of evolutionary change can be determined in broad outline in terms only of millions of years for certain parts of the horse series. It is to be hoped that in the future, the palaeontologist will be able to increase his contribution to our knowledge of organic evolution by more precise knowledge of the speed at which evolutionary change has progressed in many groups of organisms, and particularly at specific level.

THE PRE-CAMBRIAN

Nearly eight-ninths of the duration of geological time had elapsed before the earliest abundantly fossiliferous rocks were deposited. This striking fact could not be appreciated until methods for the absolute dating of rocks had been properly developed.

It seems probable that in the future, the greatest contributions made to geological knowledge by the development of geochronology will lie in the study of the Pre-Cambrian strata. It is the only known method by which we can hope to erect a world chronology of events that took place thousands of millions of years ago.

An adequate knowledge of the age relations and geographical extent of the numerous orogenic episodes which have affected the Pre-Cambrian rocks is essential before we can obtain a clear picture of the ways in which the continental masses have been built up. Have they been formed, as Tuzo Wilson believes, by a process of accretion? He pictures a succession of geosynclines that developed round the margins of the continental nuclei; each orogenic episode would weld the rocks deposited in the preceding geosyncline to the nucleus. At present we know the time relations and extent of the orogenies of the Pre-Cambrian in only broadest outline and for certain of the continents. The data at present available for the Canadian (Fig. 33) and the Baltic (Fig. 34) Shields suggests that Wilson's picture may be an oversimplification.

We do not know whether the geological processes – the rate of weathering, the rate of sedimentation, the intensity of igneous and metamorphic activity – have acted at significantly different rates during the geological past, especially in earlier Pre-Cambrian times as compared with the Phanerozoic. This has been a vexed question for many years, and until it is answered we cannot be sure to what extent the Doctrine of Uniformitarianism (p. 8) can be invoked to explain geological phenomena. It may well be misleading to interpret the past in terms of a present that is not truly representative of much of the past.

RELATIVE MOVEMENTS OF THE CONTINENTS

The development of palaeomagnetic techniques during the past few years has provided another tool for determining whether or not the continents have always been in the same position relative to one another. As with isotopic methods, one is faced with the fact that this method can be applied only to certain rock-types – in this case mainly iron-cemented sandstones and basic igneous rocks. The results already obtained provide clear evidence of the opening of the Atlantic Ocean since early Mesozoic times due to the movement apart of the American and Euro-African continents.

This is another field of geology in which quantitative data may be available when we know more precisely the exact position of the continental masses relative to one another at definite (absolute) times. We shall then know the rate of 'drifting' not only on both sides of the Atlantic, but of India away from Africa, of Australia, etc. As the Atlantic opened, the Mid-Atlantic Ridge (really a great submerged rift flanked with volcanoes) arose. Palaeomagnetic studies have shown that during the past few million years the Earth's magnetic field has periodically reversed its direction. The cause of this switch of the north and south poles is unknown. As shown in Fig. 39, these periods of normal and reversed magnetism are of very different duration when plotted against K/Ar age-determinations. They therefore promise to provide another time-scale of world-wide application, applicable to basic igneous rocks on the continents, to basic igneous rocks dredged from the ocean floors, especially from the flanks of the mid-oceanic ridges and to cores of sediments obtained from the ocean floors. Comparison of the patterns of magnetic anomalies on either side of the Mid-Atlantic Ridge indicates a rate of crustal spreading of

AGE   IN   MILLIONS   OF   YEARS

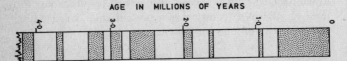

Fig. 39. *Time-scale based on direction of Earth's magnetic field. Periods of normal magnetism stippled, of reversed magnetism left blank. (After Cox, Doell and Dalrymple, 1968.)*

between one and five cm per year. There are a number of other undecided problems of major world features, which will benefit when more quantitative data, based partly on an absolute geochronology become available.

PRACTICAL VALUE

An absolute time-scale may seem, at first sight, only of academic interest. But this is a mistake. So often in geology has it been found that studies of apparently the most recondite nature have direct practical applications. The more we know concerning the make-up of the Earth's crust, the more likely we are to be able to use this knowledge in deciding the mode of formation and, therefore, the most likely places of occurrence, of materials of economic importance.

This is already happening in connection with mineralization. Isotopic dating suggests that this may have been spread over a larger period of time than was previously thought to be the case. The mineralization of southwestern England was regarded as associated with the Variscan Orogeny in late Carboniferous – early Permian times. But recent Pb/U investigations of pitchblendes have indicated that there were at least three periods of uranium mineralization at about 290 (Variscan), 225 (? early Triassic) and 50 (Lower Tertiary) million years ago. The time-relations between mineralization and igneous activity are clearly more complex than was formerly thought. Investigations on this must throw light on the vexed question as to the place of origin, probably the mantle, of the ore minerals, and the changes which they may have undergone as they moved through the rocks of the crust to the places where they are now to be found. Knowledge of such processes will be a great value in guiding the search for new minerals.

Further Reading

DUNBAR, C. O. 1966. *The Earth*. Weidenfeld and Nicolson, London.
> An attractively written and well-illustrated account of our knowledge of the nature of the Earth, its history and its place in the Solar System.

HAMILTON, E. I. 1965. *Applied geochronology*. Academic Press, London.
> Describes in detail the different methods used in isotopic age-determinations.

HARLAND, W. B. *et. al.* (Eds.) 1964. *The Phanerozoic time-scale*. Geological Society, London.
> The definitive work. A number of critical articles, followed by details of the 337 age-determinations on which the time-scale is based.

MOORBATH, S. 1967. Recent advances in the application and interpretation of radiometric age data. *Earth Science Reviews*, 3, 111–33.
> A balanced assessment of the present position.

MOORBATH, J. 1971. 'Measuring geological time' in *Understanding the Earth*, Artemus Press for the Open University, 41–52.

SHOTTON, F. W. 1967. The problems and contributions of methods of absolute dating within the Pleistocene period. *Quart. J. Geol. Soc. London*, 122, 357–84.
> Discusses in particular the reliability of C^{14} dates.

WELLS, A. K. and KIRKALDY, J. F. 1966. *Outlines of historical geology* 5th edn. Allen and Unwin, London.
> The geological history of the British Isles treated in considerable detail.

WOODFORD, A. O. 1965. *Historical geology*. W. H. Freeman, London.
> Primarily an American textbook, but deals with the other continents as well. Well illustrated and gives a good account of the successive faunas and floras.

ZEUNER, F. E. 1958. *Dating the past*. 4th edn. Methuen, London.
> First published in 1945, before the rise of isotopic dating, this book is now rather outdated, but it gives a different approach to the problems of geochronology. It is mainly concerned with the Pleistocene and the Chronology of man.

Index